MAKING MALCOLM

MAKING MALCOLM

The Myth and Meaning of Malcolm X

MICHAEL ERIC DYSON

New York   Oxford
OXFORD UNIVERSITY PRESS
1995

Oxford University Press

Oxford   New York   Toronto
Delhi   Bombay   Calcutta   Madras   Karachi
Kuala Lumpur   Singapore   Hong Kong   Tokyo
Nairobi   Dar es Salaam   Cape Town
Melbourne   Auckland   Madrid

and associated companies in
Berlin   Ibadan

Copyright © 1995 by Michael Eric Dyson

Published by Oxford University Press, Inc.,
200 Madison Avenue, New York, New York 10016

Oxford is a registered trademark of Oxford University Press

Library of Congress Cataloging-in-Publication Data
Dyson, Michael Eric.
Making Malcolm : the myth and meaning of Malcolm X /
Michael Eric Dyson.
p.   cm.
ISBN 0-19-509235-X
1. X, Malcolm, 1925–1965.   I. Title.
BP223.Z8L573338   1994   320.5'4'092—dc20   94-16396

X by Amiri Baraka, Copyright © by Amiri Baraka. Reprinted by permission
of Sterling Lord Literistic, Inc.

"What I'm Telling You" by Elizabeth Alexander, Copyright © by Elizabeth
Alexander. Reprinted by permission of the author.

9 8 7 6 5 4 3 2

Printed in the United States of America
on acid-free paper

For my marvelous son Michael Eric Dyson, II,
who, like Malcolm, is tall and intelligent
and who keeps the spirit of Malcolm alive
in his willingness to question

PREFACE
TEACHING MALCOLM

> We don't judge a man because of the color of his skin.
> We don't judge you because you're white; we don't
> judge you because you're black; we don't judge you
> because you're brown. We judge you because of what
> you do and what you practice.
>
> Malcolm X, in *Malcolm X: The Last Speeches*

"Professor Dyson, what you did was wrong, man," my student bellowed at me across my desk, supported by three of his equally angry classmates.

"You embarrassed us in front of all those people."

"No! What y'all did was wrong," I fired back at him. "You embarrassed and dissed me in front of the whole class."

I had just stormed out of the classroom and up the stairs to my space in the bank of second-floor offices in Churchill House, the building that houses Afro-American Studies at Brown, with my four accusers chasing closely behind. A full hour and a half before my seminar on Malcolm X was due to end for the day, I had publicly scolded these students—all young, black, male, and very bright—for what I felt was intolerable behavior.

"I will not put up with this any more," I angrily announced as I interrupted the class on the heels of a remark from one of the young men that proved to be the final straw.

"You have held this class hostage to your narrow beliefs and rude behavior from the very beginning. And since you've made a big deal of being a black male in this class, I resent how you've treated me—a black man—with great disrespect in my own course. Class is over."

This had never happened to me before in all my teaching career, not at Hartford Seminary, not at Chicago Theological Seminary, and not at Brown, though I had faced tough situations in each place. I felt a mixture of embarrassment and anger because I'd lost control of the class and myself, something I hadn't done even when I endured bruising battles as a pastor of three Baptist churches in the South. But I felt I had no choice.

The conflict had been building to an inevitable climax the entire semester. I knew that a seminar on a figure as explosively controversial as Malcolm X would mimic disagreements about him in other quarters of American culture. Like fiercely contested discussions about Malcolm in the popular press and in scholarly books, a seminar on Malcolm would provoke heated exchanges between students as they debated his intellectual meaning and cultural significance for black people, and for a nation that misunderstood and often reviled him when he was alive.

This proved to be true when I taught the seminar at

Brown during the previous semester. My students engaged in intense scrutiny of Malcolm's writings and pored over the secondary literature that addressed his thought and career. There were representatives from across the ideological, political, and racial spectrum in that seminar of twenty students. There were Asians and Latinas, conservatives and radicals, blacks and whites, men and women. There were black neonationalists as well as more moderate integrationists. There were students who deployed densely laden theoretical perspectives popular in post structuralism seated next to students whose unfamiliarity with the intricacies of race and class marked their beginner's pace. But the class worked because all of us fought, sometimes with ferocious abandon, sometimes with subtle restraint, over Malcolm's meaning in an environment that encouraged intellectual rigor and respect for the cantankerous differences that abounded.

Ironically, the success of my first Malcolm X seminar at Brown made me vulnerable to the sorts of problems that plagued its second incarnation a mere semester later. From the first day of class, it became obvious that this seminar possessed a dynamic that razed any arguments about race not rooted in personal experience or black cultural authority. Each class became a highly ritualized installment in a morality play about the tension between good and evil. In the case of my seminar, this conflict was crudely reduced to the tension between black and white, and, in other instances, between bourgeois and ghetto (read "authentically black") cultural and political expressions.

This narrowly conceived and troubling vision of authentic racial discourse reflects the character of debates about race now taking place in a country reeling from the implications of multiculturalism, ethnic and sexual difference, and identity politics. Often the proponents of narrow notions of racial, ethnic, or sexual identity are simply responding to the egregious and increasingly unremarked-on offenses heaped on them in a culture that remains, in crucial ways, uncomfortable with their heightened visibility. To make matters worse, blacks, gays and lesbians, Latinos and Latinas, feminists, the ghetto poor, and other minorities who defend themselves in public are often attacked under the banner of crusades against "political correctness."

"P.C." has become the common rallying cry of conservatives, liberals, and radicals, many of whom harbor resentments against the assertive presence and practices of formerly excluded minorities who no longer need representation by proxy. When these minorities show up to speak for themselves, and often in terms that are radically different from how stereotyping or scapegoating has made them appear, there is resistance from friend and foe alike. The plausible complaints that can be launched against debilitating conceptions of racial, sexual, or political identity (after all, the term "political correctness" was invented by the left to get its own house in order) are lost as anti–P.C. forces lose a sense of proportion. By and large, racial, sexual, and political minorities don't control resources or wield power in ways comparable to those used by the people and powers they oppose.

This doesn't mean that the legitimate battles that some minorities fight are not occasionally fraught with self-defeating tactics of defense that trump their highest aims. This was quite evident in my Malcolm X seminar during that second semester. It showed innocuously in small matters like the patterns of white and black bodies grouped by race in the classroom. But it was revealed more menacingly in the rigid racial reasoning of several black males who appealed to Malcolm's masculinity, his blackness, and his ghetto grounding as the basis for their strict identification with him. In their eyes, such a strategy lent their interpretations of the leader a natural advantage.

The unyielding insistence of my black male students that a racial litmus test be evoked through highly charged personal narrative (the sort of jockeying for privilege of the "because I'm black, poor, male and angry I understand him better than you" variety that I wanted to avoid) made class time a wearying exercise in either defending or defeating racial borders. The examination of ideological justifications for racist practices, or an in-depth investigation of the links between racial and class oppression—in short, the kinds of analysis that undoubtedly would have contributed to their cause—was constantly put off by appeals to a severely limiting politics of experience and authenticity.

Now I am not one of those black intellectuals who argues for a strategic point of Archimedean objectivity beyond the realm of slashing ideological conflict, longing wistfully for a neutral zone free of the fracturing results of racial politics. Neither am I committed to the belief that if Americans simply understood more about one another we could do better or that through knowledge alone, we could achieve a racial nirvana signalled in Rodney King's desperate plea that "we all get along." I understand that so much of our nation's claims of racial peace are haunted by the hypocrisy of hidden white power and concealed bigotry. I know that a great deal of discourse about race is trapped in abstraction and avoidance as we ingeniously seek to deflect the link between past injustices and present injuries.

But I also know that the game of proving you're blacker than the next Negro, an art engaged in and encouraged by Malcolm at various stages of his career, can have disastrous results. Not only does it set in motion a rancid Aristotelian regression back to a mythic "real blackness" that spins on endlessly, but it borrows from a peculiarly European quest for racial purity, a troubling Manichaean mind-set that distinguishes between "us" and "them." This is precisely the sort of thinking that anyone paying careful attention to the complex workings of black culture in Africa and throughout the diaspora cannot help but abhor.

Moreover, Malcolm's cultural renaissance—his improbable second coming—brims with irony. Our era is marked by vigorous debates about racial authenticity and selling-out, and about the consequences of crossing over to larger markets to increase mass appeal. Participants in these debates, which include not only my seminar, but everyone from politicians to rap artists, often draw on Malcolm's scorching rebukes to such moves. Meanwhile, Malcolm's X is marketed in countless business endeavors and is stylishly branded on baseball hats and T-shirts by every age, race, and gender. So much for the politics of purity.

But no matter how powerfully or with how many different examples I plied many of my black male students, they remained suspicious of any attempt to divorce Malcolm from his exclusive meaning for young, black, angry males. In many ways, they were fighting over Malcolm's tall body and short life, allowing no dibs on a legacy they felt Malcolm had bequeathed to them alone. They seemed to be saying, "White folk have ripped off so much of black culture; they can't have Malcolm too." Their black neonationalist politics led them to lay claim to Malcolm's mantle, and Malcolm's moral authority supplied support for their gestures of reappropriating his image.

Malcolm's moral authority was fueled by a moral magnetism so great that it continues to attract people who were not yet born when he met his gruesome death. Malcolm possessed an unperturbable quality that my students found irresistible: he fixed his sight on the racial goals to be obtained and pursued them with unvarying zeal. This single-mindedness is especially true of the period when, under the spell of Nation of Islam leader Elijah Muhammad's scheme of evil by colors, he pummeled the notion of black cultural inferiority with consummate skill. He eventually grew suspicious of the conclusions about race that insular versions of nationalism entail. He also gained distance on the indiscriminate demonizing of color that had already imposed severe penalties on black people's lives. He achieved a moral maturity that gave him a more complex view of the possibilities of human community across the racial divide.

Admittedly, it is difficult to highlight Malcolm's moral maturity in the midst of the contemporary reemergence of racism, a fact that many of my black male students were right to point out. In large part, Malcolm's renewed presence on the cultural scene has been made possible by students like them and by others who've kept the flame of Malcolm's life burning brightly in black nationalist book stores and community gatherings across the nation while Malcolm was out of vogue. Such groups have held up Malcolm's Promethean accomplishments of mind and body—which constitute a style of black leadership barely glimpsed in our age—as models of black cultural achievement.

More than that, Malcolm's moral authority remains intact because he wasn't given to the kind of racial hustling that often passes for leadership. No, he wasn't against using the hustling tactics he gleaned from his years on the street to portray the crimes of white racism. He knew how to rhetorically master black and white opponents. But he rarely allowed that hustling ethic to distort his relations to the black people he selflessly loved. He rarely lost sight of the fact that he was the servant, not the master, of their best hopes and bravest dreams. Malcolm's love of black people made my black male students love him in return.

But they weren't willing to subject Malcolm, the object of their devotion, to rigorous criticism. Uncritical devotion cuts across Malcolm's mature skepticism about blindly following leaders. Malcolm learned that bitter lesson in the deadly fallout of his dissociation from Elijah Muhammad, whom he had once passionately adored. Malcolm wasn't perfect, and he knew it. And those of us who want to capture some of Malcolm's magic—and to learn through critical analysis of his failures and successes, his weaknesses and strengths—don't have to make him an idol.

Malcolm's moral authority finally consisted in telling the truth about our nation as best he could. He damned its moral hypocrisy and insincerity in trying to aid the people it had harmed for so long, a fact that created seething pockets of rage within the corporate black psyche. Malcolm blessed our rage by releasing it. His tall body was a vessel for our outrage at the way things were and always had been for most black people, especially those punished by poverty and forced to live in enclaves of urban terror. It was their desire to reclaim Malcolm for themselves and to display their superior grasp of black culture that motivated many of my black male students to take my Malcolm X seminar, as I was to later discover in our confrontation in my office.

"We get tired of going to classes and having white students discuss things they don't know about, while we take a back seat and remain quiet," one of my students told me.

"When we came in the first day and saw how many white students were signed up for this class, we had a decision to make," another related to me. "We could either do like we usually do, and just say 'forget it,' because they weren't going to understand us anyway, or *we* could take charge and be the ones to set the pace."

Although I could figure out why they didn't want to let the white students have Malcolm, I found it off-putting and a bit bizarre, maybe even disingenuous, that their efforts excluded me too. After all, if the best chance of understanding Malcolm, as they often reminded me, depended on the possibility of having as close to an experience of life as Malcolm had had, then I certainly qualified, perhaps even more than they did.

I was born in 1958 in Detroit's inner city, my age, ghetto experience, and geographic proximity to his childhood digs giving me a link to Malcolm's life not immediately suggested by my students' experiences. Like Malcolm, I was reared in a large clan. My immediate family included my parents and four brothers, plus four older step-siblings who lived on their own. My father was a factory laborer; my mother, a para-professional with the Detroit public schools. Neither had been to college. My father was laid off from his job as master set-up man at Kelsey Hayes Wheel-brake and Drum Company after thirty-three years of faithful service.

After that, he started Dyson and Sons grass-cutting and sodding business, with three of us boys working with him. That work, plus our work at Morton's Nursery every day after I got home from school in the seventh grade, and later at Sam's Drugstore—besides foraging the city's alleys in search of discarded steel and iron we could turn in at the junk yard for a modest sum of money—kept us from fully going on welfare. (We did receive food stamps at one time.) Money was tight, and times were tough.

Although I received a scholarship to Cranbrook, a prestigious secondary school thirty miles from Detroit in the rich suburb of Bloomfield Hills, I left after two years and received a diploma from "night school" at Northwestern High School. Shortly after graduation at eighteen, I got my twenty-six-year-old girlfriend pregnant, and married her before our son was born in 1978. I worked at a variety of jobs (and at one time, two full-time jobs simultaneously), from fast-food to maintenance jobs, from construction work to hustling painting gigs, before I was fired from a job at Chrysler one month before my son was born. My wife and I had no insurance to cover his birth. Welfare was our only option. And as with many shotgun marriages plagued by poverty, ours misfired. We divorced scarcely a year after we'd been married.

I didn't go south to attend college until I was twenty-one. By then, I had become a Baptist minister, and I pastored three churches, and worked in a factory in Knoxville, Tennessee, to put myself through school. I graduated magna cum laude from Carson-Newman College (the first person in my family to graduate from college) and then received an M.A. and a Ph.D. in religion from Princeton University. By the time I sat before my students at Brown, I not only had been held up twice at gunpoint, but also had been a veteran of countless battles against racial, gender, and class oppression. Also I'd been virtually the sole financial supporter of my younger brother Everett's attempt to free himself from a life sentence in prison for second-degree murder. I was no stranger to young, black, angry males. I had been one, and depending on who's consulted, I'm still considered in the running. And yet, as I was to learn in our heated conflict in my office, these black males were suspicious of me from the very beginning.

"We wondered why you were teaching this course," one of the young men said, relaying a conversation they'd had earlier. "We wanted to know what your motive was for teaching a course on Malcolm." Having been a poor, angry black male who was also a preacher and from the same state where Malcolm spent his childhood wasn't enough, I guess. But I also knew that many of the young black males who were suspicious of me were greatly exaggerating their homeboy-from-the-hood backgrounds; most of them were well-to-do kids from upper-middle-class homes.

Their suspicions obviously extended to other members of the class. Time and again, in ways that were sometimes subtle, sometimes painfully conspicuous, the black males drew boundaries around Malcolm that kept even black females at bay. Sure, in the scheme of things black women had a greater chance of understanding Malcolm than, say, white men and women, but their gender prevented their complete comprehension. The hierarchy of interpretation the black males established, at once laughable and lamentable, only provoked further feelings of injury in the rest of the class.

Throughout the semester, several class members—black and white women and men, Asian and mixed-race students—came to me to vent their frustrations about many of my black male students. I felt awkward in hearing their complaints, and mostly agreeing with them. I understood what my black male students were up to, even though we were at cross-purposes as to what to make of the seminar's weekly three hours. I even gently chided the white students who came to me in private to express their bewilderment.

"I understand your feelings, even empathize with them," I told a white female student. "But we can't have a class on Malcolm X and not expect to feel some of the heat and passion he generated. The anger of some of these young black men mirrors the anger of Malcolm X."

"You're right, Professor Dyson," she responded. "I agree with you. But some of these men think it's all right to sleep with white women—'cuz some of them have slept with white women *in* our seminar. But when we make statements in class, even statements that might support Malcolm X, they frown at us."

I was sorely tempted to divulge this bit of information in the next seminar, all in the interest, mind you, of making the greatest amount of data available as we debated the precise role of experience in racial politics. But alas, cooler instincts prevailed. (My student's story reminded me of a humorous exchange between Jesse Jackson and some militant black nationalists who were criticizing him as a sellout. As he pointed out that they all had white girlfriends, they informed Jackson, "We're punishing their fathers.")

I detected, too, in the classroom exchanges between me and the black males who trailed me to my office, a generational rift whose accusing distance I had, I believed until then, successfully overcome.

"We're talking about Onyx, about 'Baccdafuccup,' " one of the black men said the first day of class, referring to a rap group and its album that had garnered critical praise for its street-savvy lyrics and its barely tempered rage. "That's who expresses what we're talking about."

Of course, there were more gentle signs of a perception

that I was no longer "young," such as the time I recently testified before a U.S. Senate subcommittee hearing on gangsta' rap. I quoted verbatim, from memory, the lyrics to a song by rapper Snoop Doggy Dogg while making a complex defense of gangsta' rappers, even as I scored their misogyny and homophobia.

"My, man," beamed a young black male in the audience who sought me out after the hearings that day, his hand extended to me in a gesture of brotherly affirmation.

"For a guy your age, you really can flow."

His compliment reminded me that while my thirty-five years made me a young man in the academy, it granted me outsider "adult" status with many young blacks.

But I also knew that more than generational forces were at work, more than age separated me from my younger compatriots. In fact, I believed that, based on his moral perspective, Malcolm X most likely would have disdained rap's materialistic impulses to get paid, spurned its hedonistic joie de vivre, its celebration of vulgar verbal expression. Hip-hop's sexual and rhetorical liberation might well have been Malcolm's moral chaos. But such dissonances between Malcolm and my young black male students went largely unnoticed, even as a basis for criticizing Malcolm's black moral puritanism, a force that surely challenged their way of life.

Malcolm's further contradictions, and the conflicting uses to which his legacy can be put, came home to me as I traveled during Thanksgiving recess to lecture on Malcolm in the Netherlands. After my public presentation, I was engaged in debate by a renowned Dutch writer, Stephan Sanders, who is black, is openly gay, and was reared in a white Dutch Jewish family. Sanders insisted that Malcolm was "far more American than he wanted to acknowledge," a product of an American culture that was obsessed with racial purity.

Sanders contended that Malcolm's lifelong struggle against domination showed classic signs of the "colonial dilemma." "The real dilemma is that being a black American as Malcolm called himself, [means that] he was born an American. So America is a . . . part of him." For Sanders, the "question is how one can definitely free oneself from the colonizer [while recognizing] that the colonizer is within you." His observations about Malcolm's internal struggles for a liberated consciousness were poignant, even moving.

Sanders couldn't understand why I voiced critical support for Malcolm, since it seemed that, as an opponent of racial essentialism, homophobia, and ethnic bigotry, I was in direct conflict with Malcolm's values. I responded that Malcolm was a complex figure, that his thought evolved, and that his moral vision was transformed over the course of a complicated, heroic life. Without either romanticizing Malcolm or making his memory a mere metaphor of rage, and thereby softening his palpable threat to black and white defenders of the status quo, I attempted to argue Malcolm's use in a progressive politics that was racially *conscious,* but not racially *exclusive.* The mature Malcolm, I contended, was open to a range of political negotiations of identity and ideology that promoted lasting liberation. That Malcolm, I claimed, was a figure who could be celebrated and put to use internationally by despised and degraded people, even as his radical edge was being rounded off by worldwide commercialization.

I felt a deep gratification in communicating the meaning of Malcolm X to an international audience in a way that I had been prevented from doing closer to home. Maybe it was that sense of foreign appreciation that helped emphasize, perhaps even exaggerate, the struggles for clarity of Malcolm's meaning I experienced back at Brown. It was not long after my return home that my class exploded and that most of my students came to me that night, heaving a collective sigh of relief, saying that they'd wished that I had dramatically confronted many of my black male students earlier.

I couldn't derive much consolation from their sentiments, as much as I personally benefitted from hearing them. I knew that many of the fears that my black males harbored—that Malcolm would be done in by sell-out Negroes, that his sometimes harsh words would be soft-pedaled to suit the crossover ambitions of people pimping off of Malcolm's newfound popularity—had already been ominously realized. In my office after the seminar's abrupt dismissal, I warned my black male students that too often conclusions about who is deemed "in" or "out" of black cultural style are based on flimsy evidence, on slippery surfaces of judgment that don't account for complexity of belief or depth of commitment.

But I also assured them that my opposition to their tendentious arguments and uselessly divisive techniques in class grew out of a deep love for them. With tears streaming down my cheeks, I confessed that I pushed them hard, yes, perhaps harder than my white students, because I expected more from them. If they were to be fearless warriors in the battle against racial oppression, then they must be prepared to wage fierce *intellectual* combat that was rooted in persuasive argumentation as well as edifying passion.

"We didn't realize that you felt this way," one of them offered as consolation to my obvious distress. "We didn't know that when you did what you did in class that you cared for us."

"Of course I care for you," I said. "You're my brothers. You're precious to me. You're precious to our race. You all have brilliant futures."

"Well, if we did anything to embarrass you, we're sorry," another replied. "We were only trying to defend ourselves."

"And if I did anything to hurt or embarrass you, I'm truly sorry," I offered in return. After this, I embraced each one of them, and we parted with the mutual benediction to "stay strong."

I am not suggesting by this seemingly sappy ending that my black male students and I don't have profound disagreements about Malcolm's meaning for our people or nation. We do. Neither do I mean that a late-evening embrace and show of solidarity between me and my black male students wiped away the hurt feelings that either side may have endured in the skirmishes that occurred over much of the seminar. It probably didn't.

What that encounter did accomplish, however, is a renewed determination on my part to make Malcolm available to the wider audience that he deserves without making him a puppet for moderate, mainstream purposes, and without freighting him with the early bigotries and blindnesses he grew to discard. Ironically, my black male students' abrasive resistance helped me understand the urgency of such a task. All of us who have a stake in the meaning and myth of Malcolm's life will continue to do battle, will continue to disagree about how and why Malcolm's memory is used in one way or another, even as we make wildly different uses of his career. From filmmakers to intellectuals, from hip-hop artists to community organizers, we are together involved in the process of exploring and evaluating the making of Malcolm's legacy. This book is a contribution to that enterprise.

Providence   M.E.D.
May 19, 1994 (Malcolm's birthday)

ACKNOWLEDGMENTS

I would like to thank my wonderful editor, Liz Maguire, for bringing this book to Oxford and for making its production such sheer pleasure. Her support of the book and its author is largely responsible for its appearance. I would also like to thank Elda Rotor and the other folk at Oxford who assisted with the book for their tireless energy in assisting Liz and me on this project. I would also like to thank my wonderful children, Michael, Mwata, and Maisha, for their love and support. I am grateful to my mother, Addie Mae Dyson, for her continued love and devotion, and to my brothers Anthony, Gregory, and Brian for their interest in my work. For my brother Everett, prisoner 212687, I am grateful for his intelligent conversation and his will to be free; stay strong and keep the faith! I am grateful to my thoughtful niece and nephew, Mejai and Torkwase Dyson, for their stimulating conversation. I am also thankful for the extremely helpful critical comments of William Van Deburg and Robin D. G. Kelley. And for my precious friend D. Soyini Madison, I am grateful for intellectual and spiritual companionship. And, of course, for my intelligent, beautiful wife, Marcia, I reserve

Acknowledgments

special gratitude for her inexhaustible store of loyalty and love, and for her belief in me and this book.

Portions of this manuscript have appeared in much different form than published here in *Social Text, Tikkun,* and *Christian Century.*

CONTENTS

MAKING MALCOLM

1
MEETING MALCOLM

> First, I don't profess to be anybody's leader. I'm one of
> 22 million Afro-Americans, all of whom have suffered
> the same things. And I probably cry out a little louder
> against the suffering than most others and therefore,
> perhaps, I'm better known. I don't profess to have a
> political, economic, or social solution to a problem as
> complicated as the one which our people face in the
> States, but I am one of those who is willing to try *any
> means necessary* to bring an end to the injustices that
> our people suffer.
>
> Malcolm X, in *By Any Means Necessary: Speeches,
> Interviews, and a Letter, by Malcolm X*

Malcolm X, one of the most complex and enigmatic African-
American leaders ever, was born Malcolm Little on May 19,
1925, in Omaha, Nebraska. Since his death in 1965, Malcolm's life has increasingly acquired mythic stature. Along
with Martin Luther King, Jr., Malcolm is a member of the
pantheon of twentieth-century black saints. Unlike that of
King, however, Malcolm's heroic rise was both aided and
complicated by his championing of black nationalism and
his advocacy of black self-defense against white racist violence.

Malcolm's ideas of black nationalism were shaped virtually from the womb by the example of his parents, Earl

and Louise Little, both members of Marcus Garvey's Universal Negro Improvement Association (UNIA). As president of the Omaha branch of the UNIA, Earl Little, who was also an itinerant Baptist preacher, vigorously proclaimed the Garveyite doctrine of racial self-help and black unity, often with Malcolm at his side. Louise Little served as reporter of the Omaha UNIA. A native of Grenada, Louise was a deeply spiritual woman who presided over her brood of eight children even as she endured the abuse of her husband, and together they heaped domestic violence on their children.

According to Malcolm, his family was driven from Omaha by the Ku Klux Klan while he was still an infant, forcing them to seek safer habitation in Lansing, the capital city of Michigan eighty miles northwest of Detroit. Their respite was only temporary, however; the Little family house was burned down by a white hate group, the Black Legionnaires, during Malcolm's early childhood in 1929. This experience of racial violence, which Malcolm termed his "earliest vivid memory," deeply influenced his unsparing denunciation of white racism during his public career as a black nationalist leader.

When Malcolm was only six, his father died after being crushed under a streetcar. It is unclear whether Earl died at the hands of the Black Legionnaires, as Malcolm reports in his autobiography, or whether his death was accidental, as recent scholarship has suggested.[1] In either case, his loss bore fateful consequences for the Little family because Louise Little was faced with raising eight children alone during the Great Depression. She eventually suffered a mental breakdown, and her children were dispersed to different foster homes.

Malcolm's life after his family's breakup went from bleak to desperate, as he was shuttled between several foster homes. Malcolm stole food to survive and began developing hustling habits that he later perfected in Boston, where he went to live with his half-sister Ella after dropping out of school in Lansing after completing the eighth grade. Before leaving school, Malcolm had become eighth-grade class president at Mason Junior High School. But a devastating rebuff from a teacher—who discouraged Malcolm in his desire to become an attorney by claiming that it was an unrealistic goal for "niggers"—finally sealed Malcolm's early fate as an academic failure.

It was in Boston that Malcolm encountered for the first time the black bourgeoisie, with its social pretensions and exaggerated rituals of cultural self-affirmation, leading him to conclude later that the black middle class was largely ineffective in achieving authentic black liberation. It was also in Boston's Roxbury and New York's Harlem that Malcolm was introduced to the street life of the northern urban poor and working class, gaining crucial insight about the cultural styles, social sufferings, and personal aspirations of everyday black people. Malcolm's hustling repertoire ranged from drug dealing and numbers running to burglary, the last activity landing him in a penitentiary for a six- to ten-year sentence. Malcolm's prison period—lasting from 1946 to 1952—marked the first of several extraordinary transformations he underwent as he searched for the truth about himself and his relation to black consciousness, black freedom and unity, and black religion.

While in prison, Malcolm read widely and argued passionately about a broad scope of subjects, from biblical theology to Western philosophy, voraciously absorbing the work of authors as diverse as Louis S. B. Leakey and Friedrich Nietzsche. Malcolm read so much during this period that his eyesight became strained, and he began wearing his trademark glasses. It was during his prison stay that Malcolm experienced his first religious conversion, slowly evolving from a slick street hustler and con artist to a sophisticated, self-taught devotee of Elijah Muhammad and the Nation of Islam, the black nationalist religious group that Muhammad headed. Malcolm was drawn to the Nation of Islam because of the character of its black nationalist practices and beliefs: its peculiar gift for rehabilitating black male prisoners; its strong emphasis on black pride, history, culture, and unity; and its unblinking assertion that white men were devils, a belief that led Muhammad and his followers to advocate black separation from white society.

Within a year of his release from prison on parole in 1952, Malcolm became a minister with the Nation of Islam, journeying to its Chicago headquarters to meet face to face with the man whose theological doctrines of white evil and black racial superiority had given Malcolm new life. Through a herculean work ethic and spartan self-discipline—key features of the black puritanism that characterized the Nation's moral orientation—Malcolm worked his way in short order from assistant minister of Detroit's Temple Number One to national spokesman for Elijah Muhammad and the Nation of Islam. In his role as the mouthpiece for the Nation of Islam, Malcolm brought unprecedented visibility to a religious group that many critics had either ignored or dismissed as fundamentalist fringe fanatics. Under Malcolm's leadership, the Nation grew from several hundred to a hundred thousand members by the early 1960s. The Nation under Malcolm also produced forty temples throughout the United States and purchased thirty radio stations.

During the late 1950s and early 1960s, enormous changes were rapidly occurring within American society in regard to race. The momentous *Brown* v. *Board of Education* Supreme Court decision, delivered in 1954, struck down the "separate-but-equal" law that had enforced racially segregated public education since 1896. And in 1955, the historic bus boycott in Montgomery, Alabama—sparked by seamstress Rosa Parks's refusal to surrender her seat to a white passenger, as legally mandated by a segregated public-transportation system—brought its leader, Martin Luther King, Jr., to national prominence. King's fusion of black

Christian civic piety and traditions of American public morality and radical democracy unleashed an irresistible force
on American politics that fundamentally altered the social
conditions of millions of blacks, especially the black middle
classes in the South.

The civil rights movement, though, barely affected the
circumstances of poor southern rural blacks. Neither did it
greatly enhance the plight of poor northern urban blacks,
whose economic status and social standing were severely
handicapped by forces of deindustrialization: the rise of au-
tomated technology that displaced human wage earners,
the severe decline in manufacturing and in retail and
wholesale trade, and escalating patterns of black unem-
ployment. These social and economic trends, coupled with
the growing spiritual despair that beginning in the early
1950s gripped Rust Belt cities like New York, Chicago, Phil-
adelphia, Detroit, Cleveland, Indianapolis, and Baltimore,
did not initially occupy the social agenda of the southern-
based civil rights movement.

Malcolm's ministry, however, as was true of the Nation
of Islam in general, was directed toward the socially dispossessed, the morally compromised, and the economically
desperate members of the black proletariat and ghetto poor
who were unaided by the civil rights movement. The Nation
of Islam recruited many of its members among the prison
populations largely forgotten by traditional Christianity
(black and white). The Nation also proselytized among the

hustlers, drug dealers, pimps, prostitutes, and thieves whose lives, the Nation contended, were ethically impoverished by white racist neglect of their most fundamental needs: the need for self-respect, the need for social dignity, the need to understand their royal black history, and the need to worship and serve a black God. All of these were provided in the black nationalist worldview of the Nation of Islam.

Malcolm's public ministry of proselytizing for the Nation of Islam depended heavily on drawing contrasts between what he and other Nation members viewed as the corruption of black culture by white Christianity (best symbolized in Martin Luther King, Jr., and segments of the civil rights movement) and the redemptive messages of racial salvation proffered by Elijah Muhammad. Malcolm relentlessly preached the virtues of black self-determination and self-defense even as he denounced the brainwashing of black people by Christian preachers like King who espoused passive strategies of resistance in the face of white racist violence.

Where King advocated redemptive suffering for blacks through their own bloodshed, Malcolm promulgated "reciprocal bleeding" for blacks and whites. As King preached the virtues of Christian love, Malcolm articulated black anger with unmitigated passion. While King urged nonviolent civil disobedience, Malcolm promoted the liberation of blacks by whatever means were necessary, including (though not exclusively, as some have argued) the possibility of armed self-defense. While King dreamed, Malcolm saw nightmares.

It was Malcolm's unique ability to narrate the prospects of black resistance at the edge of racial apocalypse that made him both exciting and threatening. Malcolm spoke out loud what many blacks secretly felt about racist white people and practices, but were afraid to acknowledge publicly. Malcolm boldly specified in lucid rhetoric the hurts, agonies, and frustrations of black people chafing from an enforced racial silence about the considerable cultural costs of white racism.

Unfortunately, as was the case with most of his black nationalist compatriots and civil rights advocates, Malcolm cast black liberation in terms of masculine self-realization. Malcolm's zealous trumpeting of the social costs of black male cultural emasculation went hand in hand with his often aggressive, occasionally vicious, put-downs of black women. These slights of black women reflected the demonology of the Nation of Islam, which not only viewed racism as an ill from outside its group, but argued that women were a lethal source of deception and seduction from within. Hence, Nation of Islam women were virtually desexualized through "modest" dress, kept under the close supervision of men, and relegated to the background while their men took center stage. Such beliefs reinforced the already inferior position of black women in black culture.

These views, ironically, placed Malcolm and the Nation of Islam squarely within misogynist traditions of white and black Christianity. It is this aspect, especially, of Malcolm's public ministry that has been adopted by contemporary black urban youth, including rappers and filmmakers. Although Malcolm would near the end of his life renounce his sometimes vitriolic denunciations of black women, his contemporary followers have not often followed suit.

But as the civil rights movement expanded its influence, Malcolm and the Nation came under increasing criticism for its deeply apolitical stance. Officially, the Nation of Islam was forbidden by Elijah Muhammad to become involved in acts of civil disobedience or social protest, ironically containing the forces of anger and rage that Malcolm's fiery rhetoric helped unleash. This ideological constraint stifled Malcolm's natural inclination to action, and increasingly caused him great discomfort as he sought to explain publicly the glaring disparity between the Nation's aggressive rhetoric and its refusal to become politically engaged.

Malcolm's growing dissatisfaction with the Nation's apolitical posture only deepened his suspicions about its leadership role in aiding blacks to achieve real liberation. Malcolm also became increasingly aware of the internal corruption of the Nation—unprincipled financial practices among top officials who reaped personal benefit at the expense of the rank and file, and extramarital affairs involving leader Elijah Muhammad. Moreover, there is evidence that Malcolm had privately forsaken his belief in the whites-are-devils doctrine years before his widely discussed public rejection of the doctrine after his 1964 split from the Nation of Islam, his embrace of orthodox Islamic belief, and his religious pilgrimage to Mecca.[2]

The official cause of Malcolm's departure from the Nation of Islam was Elijah Muhammad's public reprimand of Malcolm for his famous comments that President John F. Kennedy's assassination merely represented the "chickens coming home to roost." Malcolm was saying that the violence the United States had committed in other parts of the world was returning to haunt this nation. Muhammad quickly forbade Malcolm from publicly speaking, initially for ninety days, motivated as much by jealousy of Malcolm's enormous popularity among blacks outside the Nation of Islam as by his desire to punish Malcolm for a comment that would bring the Nation undesired negative attention from an already racially paranoid government.

In March 1964, Malcolm left the Nation of Islam after it became apparent that he could not mend his relationship with his estranged mentor. He formed two organizations, one religious (Muslim Mosque) and the other political (Organization of Afro-American Unity, or OAAU). The OAAU was modeled after the Organization of African Unity and reflected Malcolm's belief that broad social engagement provided blacks their best chance for ending racism. Before establishing the OAAU, however, Malcolm fulfilled a long-standing dream of making a hajj to Mecca. While there, Malcolm wrote a series of letters to his followers detailing his stunning change of heart about race relations, declaring that his humane treatment by white Muslims and his perception of the universality of Islamic religious truth had forced him to reject his former narrow beliefs about whites. Malcolm's change of heart, though, did not blind him to the persistence of American racism and the need to oppose its broad variety of expressions with aggressive social resistance.

After his departure from the Nation of Islam, Malcolm traveled extensively, including trips to the Middle East and Africa. His travels broadened his political perspective considerably, a fact reflected in his new appreciation of socialist movements (though he didn't embrace socialism) and a new international note in his public discourse as he emphasized the link between African-American liberation and movements for freedom throughout the world, especially in African nations. Malcolm didn't live long enough to fulfill the promise of his new directions. On February 21, 1965, three months shy of his fortieth birthday, Malcolm X was gunned down by Nation of Islam loyalists as he prepared to speak to a meeting of the OAAU. Fortunately, Malcolm had recently completed his autobiography with the help of Alex Haley. That work, *The Autobiography of Malcolm X*, stands as a classic of black letters and American autobiography.[3]

Malcolm lived only fifty weeks after his break with the Nation of Islam, initiating his last and perhaps most meaningful transformation of all: from revolutionary black nationalist to human rights advocate. Although Malcolm never gave up on black unity or self-determination—and neither did he surrender his acerbic wit on behalf of the voiceless millions of poor blacks who could never speak their pain before the world—he did expand his field of vision to include poor, dispossessed people of color from around the world, people whose plight resulted from class inequality and economic oppression as much as from racial domination. Had he lived, we can only hope that vexing contemporary problems from gender oppression to homophobia might have exercised his considerable skills of social rage and incisive, passionate oratory in giving voice to fears and resentments that most people can speak only in private.

During the last year of his life, Malcolm's social criticism and political engagement reflected a will to spontaneity, his analysis an improvisatory and fluid affair that drew from his rapidly evolving quest for the best means available for real black liberation—but a black liberation connected to the realization of human rights for all suffering peoples. In the end, Malcolm's moral pragmatism and experimental social criticism linked him more nearly to the heart of African-American culture and American radical practices than it might have otherwise appeared during his controversial career. Malcolm's complexity resists neat categories of analysis and rigid conclusions about his meaning.

It is this Malcolm—the Malcolm who spoke with uncompromising ardor about the poor, black, and dispossessed, and who named racism when and where he found it—who appealed to me as a young black male coming to maturity during the 1970s in the ghetto of Detroit. I took pleasure in his early moniker Detroit Red, feeling that our common geography joined us in a project to reclaim the dignity of black identity from the chaotic dissemblances and self-deceptions instigated by racist oppression.

But the riots of 1967—with their flames of frustration burning bitterly in my neighborhood, a testament to the unreconciled grievances that fueled racial resentments—had already confirmed Malcolm's warnings about the desperate state of urban black America. And the death of Martin Luther King, Jr., one year later ruptured the veins of nonviolent response to black suffering, evoking seizures of social unrest in the nerve centers of hundreds of black communities across the nation. King's death and Malcolm's life forced me to grapple with the best remedy for resisting racism.

As a result, I turned more frequently to a means of communication and combat that King and Malcolm had favored and that had been nurtured in me by my experience in the black church: rhetorical resistance. In African-American cultures, acts of rhetorical resistance are often more than mere words. They encompass a complex set of symbolic expressions and oral interactions with the "real" world. These expressions and interactions are usually supported by substantive black cultural traditions—from religious worship to social protest—that fuse speech and performance. Much of the ingenuity and inventiveness of black rhetorical resistance was evident in the church-based civil rights movement and in black nationalist struggles for self-determination in the 1960s.

One form of rhetorical resistance that has been prominently featured throughout black cultural history is the black sermon, the jewel in the crown of black sacred rhetoric. Here, a minister, or another authorized figure, thrives in the delivery of priestly wisdom and prophetic warning through words of encouragement and comfort, of chastening and challenge. Martin and Malcolm, of course, were widely acknowledged masters of black sacred rhetoric—as well as brilliant political rhetoricians whose deft weaving of spiritual uplift and secular complaint forged a powerful basis for black action in a bruising white world. The excellent examples of Martin and Malcolm—along with the more immediate impact of my pastor, Frederick G. Sampson—brought me to believe that words can have world-making and life-altering consequences.

In the years following Malcolm's and Martin's deaths, I participated in all manner of black public oral performance—from church plays and speeches to poetry recitations and oratorical contests—that whetted my appetite for the word. At eleven, I wrote a speech for the local Optimist Club that won me a first-place trophy and a photograph and headline in the *Detroit News* that read "Boy's Plea Against Racism Wins Award." Martin's and Malcolm's spirit hovered intimately around my performance. Their presence in word also inspired my decision to become an ordained Baptist minister, and sustained me as I became, in quick succession, a teen-age father, a welfare recipient, a wheel-brake-and-drum-factory laborer, and a pastor in the South.

As I have matured, journeying from factory worker to professor, it is the Malcolm who valued truth over habit who has appealed most to me, his ability to be self-critical and to change his direction an unfailing sign of integrity and courage. But these two Malcolms need not be in ultimate, fatal conflict, need not be fractured by the choice between seeking an empowering racial identity and linking ourselves to the truth no matter what it looks like, regardless of color, class, gender, sex, or age. They are both legitimate quests, and Malcolm's career and memory are enabling agents for both pursuits. His complexity is our gift.

PART I
MALCOLM X'S INTELLECTUAL LEGACY

If I say, my father was Betty Shabazz's lawyer, the
poem can go no further. I've given you the punchline.
If you know who she is, all you can think about is how
and what you want to know about me, about my father,
about Malcolm, especially in 1990 when he's all over t-
shirts and medallions, but what I'm telling you is that
Mrs. Shabazz was a nice lady to me, and I loved her
name for the wrong reasons, SHABAZZ! and what I
remember is going to visit her daughters in 1970 in a
dark house with little furniture and leaving with a
candy necklace the daughters gave me, to keep. Now that
children see his name and call him, Malcolm Ten, and
someone called her Mrs. Ex-es, and they don't really
remember who he was or what he said or how he smiled
the way it happened when it did, and neither do I, I
think about how history is made more than what happened
and about a nice woman in a dark house filled with
daughters and candy, something dim and unspoken,
expectation.

Elizabeth Alexander, "What I'm Telling You"

2
X MARKS THE PLOTS: A CRITICAL READING OF MALCOLM'S READERS

I think all of us should be critics of each other. When-
ever you can't stand criticism you can never grow. I
don't think that it serves any purpose for the leaders
of our people to waste their time fighting each other
needlessly. I think that we accomplish more when we
sit down in private and iron out whatever differences
that may exist and try and then do something con-
structive for the benefit of our people. But on the other
hand, I don't think that we should be above criticism.
I don't think that anyone should be above criticism.

Malcolm X, in *Malcolm X: The Last Speeches*

The life and thought of Malcolm X have traced a curious
path to black cultural authority and social acceptance
since his assassination in 1965. At the time of his mar-
tyrdom—achieved through a murder that rivaled in its
fumbling but lethal execution the treacherous twists of a
Shakespearean tragedy—Malcolm was experiencing a radical shift in the personal and political understandings that

governed his life and thought.[1] Malcolm's death heightened the confusion that had already seized his inner circle because of his last religious conversion. His death also engendered bitter disagreement among fellow travelers about his evolving political direction, conflicts that often traded on polemic, diatribe, and intolerance. Thus Malcolm's legacy was severely fragmented, his contributions shredded in ideological disputes even as ignorance and fear ensured his further denigration as the symbol of black hatred and violence.

Although broader cultural investigation of his importance has sometimes flagged, Malcolm has never disappeared among racial and political subcultures that proclaim his heroic stature because he embodied ideals of black rebellion and revolutionary social action.[2] The contemporary revival of black nationalism, in particular, has focused renewed attention on him. Indeed, he has risen to a black cultural stratosphere that was once exclusively occupied by Martin Luther King, Jr. The icons of success that mark Malcolm's ascent—ranging from posters, clothing, speeches, and endless sampling of his voice on rap recordings—attest to his achieving the pinnacle of his popularity more than a quarter century after his death.

Malcolm, however, has received nothing like the intellectual attention devoted to Martin Luther King, Jr., at least nothing equal to his cultural significance. Competing waves of uncritical celebration and vicious criticism—which settle easily into myth and caricature—have undermined appreciation of Malcolm's greatest accomplishments. The peculiar needs that idolizing or demonizing Malcolm fulfill mean that intellectuals who study him are faced with the difficult task of describing and explaining a controversial black leader and the forces that produced him.[3] Such critical studies must achieve the "thickest description" possible of Malcolm's career while avoiding explanations that either obscure or reduce the complex nature of his achievements and failures.[4]

Judging by these standards, the literature on Malcolm X has often missed the mark. Even the classic *Autobiography of Malcolm X* reflects both Malcolm's need to shape his personal history for public racial edification while bringing coherence to a radically conflicting set of life experiences and coauthor Alex Haley's political biases and ideological purposes.[5] Much writing about Malcolm has either lost its way in the murky waters of psychology dissolved from history or simply substituted—given racial politics in the United States—defensive praise for critical appraisal. At times, insights on Malcolm have been tarnished by insular ideological arguments that neither illuminate nor surprise. Malcolm X was too formidable a historic figure—the movements he led too variable and contradictory, the passion and intelligence he summoned too extraordinary and disconcerting—to be viewed through a narrow cultural prism.

My intent in this chapter is to provide a critical path through the quagmire of conflicting views of Malcolm X. I have identified at least four Malcolms who emerge in the intellectual investigations of his life and career: Malcolm as hero and saint, Malcolm as a public moralist, Malcolm as victim and vehicle of psychohistorical forces, and Malcolm as revolutionary figure judged by his career trajectory from nationalist to alleged socialist. Of course, many treatments of Malcolm's life and thought transgress rigid boundaries of interpretation. The Malcolms I have identified, and especially the categories of interpretation to which they give rise, should be viewed as handles on broader issues of ideological warfare over who Malcolm is, and to whom he rightfully belongs. In short, they help us answer Whose Malcolm is it?

I am not providing an exhaustive review of the literature, but a critical reading of the dominant tendencies in the writings on Malcolm X.[6] The writings make up an intellectual universe riddled with philosophical blindnesses and ideological constraints, filled with problematic interpretations, and sometimes brimming with brilliant insights. They reveal as much about the possibilities of understanding and explaining the life of a great black man as they do about Malcolm's life and thought.

Hero Worship and the Construction of a Black Revolutionary

In the tense and confused aftermath of Malcolm's death, several groups claiming to be his ideological heirs competed in a warfare of interpretation over Malcolm's torn legacy. The most prominent of these included black nationalist and revolutionary groups such as the Student Nonviolent Coordinating Committee (SNCC, under the leadership of Stokely Carmichael), the Congress of Racial Equality (CORE, under the leadership of James Farmer and especially Floyd McKissick), the Black Panther party, the Republic of New Africa, and the League of Revolutionary Black Workers.[7] They appealed to his vision and spirit in developing styles of moral criticism and social action aimed at the destruction of white supremacy. These groups also advocated versions of Black Power, racial self-determination, black pride, cultural autonomy, cooperative socialism, and black capitalism.[8]

Malcolm's death also caused often bitter debate between custodians of his legacy and his detractors, either side arguing his genius or evil in a potpourri of journals, books, magazines, and newspapers. For many of Malcolm's keepers, the embrace of his legacy by integrationists or Marxists out to re-create Malcolm in their distorted image was more destructive than his critics characterizing him in exclusively pejorative terms.

For all his nationalist followers, Malcolm is largely viewed as a saintly figure defending the cause of black unity while fighting racist oppression. Admittedly, the development of stories that posit black heroes and saints serves a crucial cultural and political function. Such stories may be used to combat historical amnesia and to challenge the deification of black heroes—especially those deemed capable of betraying the best interests of African-Americans—by forces outside black communities. Furthermore, such stories reveal that the creation of (black) heroes is neither accidental nor value neutral, and often serves political ends that are not defined or controlled by black communities. Even heroes proclaimed worthy of broad black support are often subject to cultural manipulation and distortion.

The most striking example of this involves Martin Luther King, Jr. Like Malcolm X, King was a complex historic figure whose moral vision and social thought evolved over time.[9] When King was alive, his efforts to effect a beloved community of racial equality were widely viewed as a threat to a stable social order. His advocacy of nonviolent civil disobedience was also viewed as a detrimental detour from the proper role that religious leaders should play in public. Of course, the rise of black radicalism during the late 1960s softened King's perception among many whites and blacks. But King's power to excite the social imagination of Americans only increased after his assassination.

The conflicting uses to which King's memory can be put—and the obscene manner in which his radical legacy can be deliberately forgotten—are displayed in aspects of the public commemoration of his birthday. To a significant degree, perceptions of King's public aims have been shaped by the corporate sector and (sometimes hostile) governmental forces. These forces may be glimpsed in Coca-Cola commercials celebrating King's birthday, and in Ronald Reagan's unseemly hints of King's personal and political defects at the signing of legislation to establish King's birthday as a national holiday.

King's legacy is viewed as most useful when promoting an unalloyed optimism about the possibilities of American social transformation, which peaked during his "I Have a Dream" speech. What is not often discussed—and is perhaps deliberately ignored—is how King dramatically revised his views, glimpsed most eloquently in his Vietnam-era antiwar rhetoric and in his War-on-Poverty social activism. Corporation-sponsored commercials that celebrate King's memory—most notably, television spots by McDonald's and Coca-Cola aimed at connecting their products to King's legacy—reveal a truncated understanding of King's meaning and value to American democracy. These and other efforts at public explanation of King's meaning portray his worth as underwriting the interests of the state, which advocates a distorted cultural history of an era actually shaped more by blood and brutality than by distant dreams.

Many events of public commemoration avoid assigning specific responsibility for opposition to King's and the civil rights movement's quest for equality. On such occasions, the uneven path to racial justice is often described in a manner that makes progress appear an inevitable fact of our national life. Little mention is made of the concerted efforts—not only of bigots and white supremacists, but, more important, of government officials and average citizens—to stop racial progress. Such stories deny King's radical challenge to narrow conceptions of American democracy. Although King and other sacrificial civil rights participants are lauded for their possession of the virtue of courage, not enough attention is given to the vicious cultural contexts that called forth such heroic action.

Most insidious of all, consent to these whitewashed stories of King and the 1960s is often secured by the veiled threat that King's memory will be either celebrated in this manner or forgotten altogether. The logic behind such a threat is premised on a belief that blacks should be grateful for the state's allowing King's celebration to occur at all. These realities make the battle over King's memory—waged by communities invested in his radical challenge to American society—a constant obligation. The battle over King's memory also provides an important example to communities interested in preserving and employing Malcolm's memory in contemporary social action. As with King, making Malcolm X a hero reveals the political utility of memory and reflects a deliberate choice made by black communities to identify and honor the principles for which Malcolm lived and died.

For many adherents, Malcolm remained until his death a revolutionary black nationalist whose exclusive interest was to combat white supremacy while fostering black unity. Although near the end of his life Malcolm displayed a broadened humanity and moral awareness—qualities overlooked by his unprincipled critics and often denied by his true believers—his revolutionary cohorts contended that Malcolm's late-life changes were cosmetic and confused, the painful evidence of ideological vertigo brought on by paranoia and exhaustion.

All these interpretations are vividly elaborated in John Henrik Clarke's anthology *Malcolm X: The Man and His Times*.[10] Clarke's book brings together essays, personal reflections, interviews, and organizational statements that provide a basis for understanding and explaining different dimensions of Malcolm's life and career. Although its various voices certainly undermine a single understanding of Malcolm's meaning as a father, leader, friend, and husband (after all, it includes writers as different as Albert Cleage and Gordon Parks), the book's tone suggests an exercise in hero worship and saint making, as cultural interpreters gather and preserve fragments of Malcolm's memory.

Thus even the power of an individual essay to critically engage an aspect of Malcolm's contribution or failure is overcome by the greater urgency of the collective enterprise: to establish Malcolm as a genuine hero of the people, but more than this, a sainted son of revolutionary struggle who was perfectly fit for the leadership task he helped define. But moments of criticism come through. For instance, in the course of a mostly favorable discussion of Malcolm's leadership, Charles Wilson insightfully addresses the structural problems confronting black protest leaders as he probes Malcolm's "failure of leadership style and a failure to evolve a sound organizational base for his activities," concluding that Malcolm was a "victim of his own charisma."[11]

At least two other writers in the collection also attempt to critically explore Malcolm's limitations and the distortions of his legacy by other groups. James Boggs deplores both the racism of white Marxist revolutionaries who cannot see beyond color and the lack of "scientific analysis" displayed by Malcolm's black nationalist heirs whose activity degenerates into Black Power sloganeering. And Wyatt Tee Walker, King's former lieutenant, criticizes Malcolm for "useless illogical and intemperate remarks that helped neither him nor his cause," while emphasizing the importance of Malcolm's pro-black rhetoric and his promulgation of the right to self-defense.[12] At the same time, Walker uselessly repeats old saws about the vices of black matriarchy.

But these flutters of criticism are mostly overridden by the celebrative and romantic impulses that are expressed in several essays. Fortunately, Patricia Robinson's paean to Malcolm X as a revolutionary figure stops short of viewing black male patriarchy as a heroic achievement. Instead, she sees Malcolm as the beginning of a redeemed black masculinity that helps, not oppresses, black children and women. But in essays by W. Keorapetse Kgositsile, Abdelwahab M. Elmessiri, and Albert Cleage, Malcolm's revolutionary black nationalist legacy is almost breathlessly, even reverentially, evoked.

Cleage especially, in his "Myths About Malcolm X," seeks to defend Malcolm's black nationalist reputation from assertions that he was becoming an integrationist, an internationalist, or a Trotskyist Marxist, concluding that "if in Mecca he had decided that blacks and whites can unite, then his life at that moment would have become meaningless in terms of the world struggle of black people."[13] Clarke's book makes sense, especially when viewed against the historical canvas of late-1960s racial politics and in light of the specific cultural needs of urban blacks confronting deepening social crisis after Malcolm's death. But its goal of redeeming Malcolm's legacy through laudatory means makes its value more curatorial than critical.

Similarly, Oba T'Shaka's *The Political Legacy of Malcolm*

X is an interpretation of Malcolm X as a revolutionary black nationalist, and *The End of White World Supremacy: Four Speeches by Malcolm X,* edited by Benjamin Karim, attempts to freeze Malcolm's development in the fateful year before his break with Elijah Muhammad and the Nation of Islam.[14] T'Shaka is an often perceptive social critic and political activist who believes that "the scattering of Africans throughout the world gave birth to the idea of Pan-Africanism," and that the "oppression of Blacks in the United States cannot be separated from the oppression of Africans on the African continent and in the world."[15]

Such an international perspective establishes links between blacks throughout the world, forged by revolutionary black nationalist activity expressed in political insurgency, material and resource sharing, and the exchange of ideas. In this context, T'Shaka maintains that Malcolm was a revolutionary black nationalist who "identified the world-wide system of white supremacy as the number one enemy of Africans and people of color throughout the world." He argues that Malcolm's internationalist perspective on revolutionary political resistance was specifically linked to African experiments in socialist politics, contending that Malcolm rejected European models of political transformation. Not surprisingly, T'Shaka is sour on the notion that after his trip to Mecca Malcolm accepted and expressed support for black–white unity, and he characterizes beliefs that Malcolm began to advocate a Trotskyite social-ism as "farfetched statements."[16]

Although he gives a close reading of Malcolm's ideas, T'Shaka's treatment of Malcolm is marred by largely un-critical explorations of Malcolm's rhetoric. He fails to chal-lenge Malcolm's philosophical presuppositions or even critically to juxtapose contradictory elements of Malcolm's rhetoric. In effect, he bestows a canonical cloak on Mal-colm's words. Nor does T'Shaka give a persuasive expla-nation of the social forces and political action that shaped Malcolm's thinking in his last years. Understanding these facts might illuminate the motivation behind Malcolm's utopian interpretations of black separatist ideology, which maintained that racial division was based on blacks pos-sessing land either in Africa or in the United States. Al-though T'Shaka, following Malcolm's own schema, draws distinctions between his long-range program (that is, return to Africa, which he claims Malcolm never gave up) and short-term tactics (that is, cultural, psychological, and phil-osophical migration), he doesn't prove that Malcolm ever resolved the ideological tensions in black nationalism.

Karim's *The End of White World Supremacy* is an attempt to wrench Malcolm's speeches from their political context and place them in a narrative framework that uses Mal-colm's own words—even after his break with the Nation of Islam—to justify Elijah Muhammad's religious theodicy.

Such a move ignores Malcolm's radically transformed self-understanding and asserts, through his own words, a worldview he eventually rejected. Karim, who as Benjamin Goodman was Malcolm's close associate through his Nation of Islam phase until his death, says in his introduction that Muhammad gave Malcolm "the keys to knowledge and understanding," that this is "one key point in Malcolm's life that is still generally misunderstood, or overlooked," and that these speeches "represent a fair cross section of his teaching during that crucial last year as a leader in the Nation of Islam."[17]

Karim's introduction to the speeches winks away the ideological warfare that helped drive Malcolm from the Nation of Islam, and ignores evidence that Malcolm grew to characterize his years with Muhammad as "the sickness and madness of those days."[18] Here we have Malcolm the master polemicist telling twice-told tales of Mr. Yacub and white devils, a doctrine he had long forsaken. Here, too, is Malcolm the skillful dogmatist deriding Paul Robeson for not knowing his history, when in reality Malcolm grew to admire Robeson and tried to meet him a month before his own death.[19] The political context Karim gives to the speeches attempts to transform interesting and essential historical artifacts from Malcolm's past into a living document of personal faith and belief.

Karim's shortcomings reveal the futility of examining

A Critical Reading of Malcolm's Readers

Malcolm's life and thought without regard for sound historical judgment and intellectual honesty. Serious engagement with Malcolm's life and thought must be critical and balanced. The most useful evaluations of Malcolm X are those anchored in forceful but fair criticism of his career that hold him to the same standards of scholarly examination as we would any figure of importance to (African-) American society. But such judgments must acknowledge the tattered history of vicious, uncomprehending, and disabling cultural criticism aimed at black life, a variety of criticism reflected in many cultural commentaries on Malcolm's life.[20]

The overwhelming weakness of hero worship, often, is the belief that the community of hero worshipers possesses the *definitive* understanding of the subject—in this instance, Malcolm—and that critical dissenters from the received view of Malcolm are traitors to black unity, inauthentic heirs to his political legacy, or misguided interpreters of his ideas.[21] This is even more reason for intellectuals to bring the full weight of their critical powers to bear on Malcolm's life. Otherwise, his real brilliance will be diminished by efforts to canonize his views without first considering them, his ultimate importance as a revolutionary figure sacrificed to celebratory claims about his historic meaning. Toward this end, Malcolm's words best describe the critical approach that should be adopted in examining his life and thought:

Now many Negroes don't like to be criticized—they don't like for it to be said that we're not ready. They say that that's a stereotype. We have assets—we have liabilities as well as assets. And until our people are able to . . . analyze ourselves and discover our own liabilities as well as our assets, we never will be able to win any struggle that we become involved in. As long as the black community and the leaders of the black community are afraid of criticism and want to classify all criticism, collective criticism, as a stereotype, no one will ever be able to pull our coat. . . . [W]e have to . . . find out where we are lacking, and what we need to replace that which we are lacking, [or] we never will be able to be successful.[22]

The Vocation of a Public Moralist

Within African-American life, a strong heritage of black leadership has relentlessly and imaginatively addressed the major obstacles to the achievement of a sacred trinity of social goods for African-Americans: freedom, justice, and equality. Racism has been historically viewed as the most lethal force to deny black Americans their share in the abundant life that these goods make possible. The central role that the church has traditionally played in many black communities means that religion has profoundly shaped the moral vision and social thought of black leaders' responses to racism.[23] Because freedom, justice, and equality have been viewed by black communities as fundamental in the exercise of citizenship rights and the expression of social dignity, a diverse group of black leaders has advocated varied models of racial transformation in public life.

The centrality of Christianity in African-American culture means that the moral character of black public protest against racism has oscillated between reformist and revolutionary models of racial transformation. From Booker T. Washington to Joseph H. Jackson, black Christian reformist approaches to racial transformation have embraced liberal notions of the importance of social stability and the legitimacy of the state. Black Christian reformist leaders have sought to shape religious resistance to oppression, inequality, and injustice around styles of rational dissent that reinforce a stable political order. From Nat Turner to the latter-day Martin Luther King, Jr., black Christian revolutionary approaches to racial transformation have often presumed the fundamental moral and social limitations of the state. Black Christian revolutionary leaders have advocated public protest against racism in a manner that disrupts the forceful alliance of unjust social privilege and political legitimacy that have undermined African-American life.

In practice, black resistance to American racism has fallen somewhere between these two poles. At their best, black leaders have opposed American racism while appealing to religion and politics in prescribing a remedy. Whether influenced by black Christianity, Black Muslim belief, or other varieties of black religious experience, proponents of public morality combined spiritual insight with political resistance in the attempt to achieve social reconstruction. Any effort to understand Malcolm X, and the cultural and religious beliefs he appealed to and argued against in making his specific claims, must take these traditions of prophetic and public morality into consideration.

Of the four books that largely view Malcolm's career through his unrelenting ethical insights and the moral abominations to which his vision forcefully responded, Louis Lomax's *When the Word Is Given: A Report on Elijah Muhammad, Malcolm X, and the Black Muslim World* and James Cone's *Martin and Malcolm and America: A Dream or a Nightmare?* treat the religious roots of Malcolm's moral vision. Peter Goldman's *The Death and Life of Malcolm X* and Lomax's *To Kill a Black Man* expound the social vision and political implications of Malcolm's moral perspective. Moreover, both Lomax's and Cone's books are comparative studies of Malcolm and Martin Luther King, Jr., Malcolm's widely perceived ideological opposite. The pairing of these figures invites inquiry about the legitimacy and usefulness of such comparisons, questions I will take up later.[24]

Lomax's *When the Word Is Given* is a perceptive and informal ethnography of the inner structure of belief of the Nation of Islam, a journalist's attempt to unveil the mysterious concatenation of religious rituals, puritanical behavior, and unorthodox beliefs that have at once intimidated and intrigued outsiders. Although other, more scholarly critics have examined Black Muslim belief, Lomax is a literate amateur whose lucid prose and imaginative reporting evoke the electricity and immediacy of the events he describes.[25]

Lomax is also insightful in his description of the cultural forces that helped bring Black Muslim faith into existence. He artfully probes how the Nation of Islam proved essential during the 1950s and 1960s for many black citizens who were vulnerably perched at the crux of the racial dilemma in the United States, seeking psychic and social refuge from the insanity of the country's fractured urban center. In Lomax's portrait, it is at the juncture between racist attack and cultural defense that Malcolm X's moral vocation emerges: he voices the aspirations of the disenfranchised, the racially displaced, the religiously confused, and the economically devastated black person. As Lomax observes, the "Black Muslims came to power during a moral interregnum"; Malcolm "brings his message of importance and dignity to a class of Negroes who have had little, if any, reason to feel proud of themselves as a race or as individuals."[26]

Despite the virtue of including several of Malcolm's speeches and interviews, which compose the second half of the book (including an interview during Malcolm's suspension from the Nation), the study's popular purposes largely stifle a sharp analysis of Malcolm's moral thought. Lomax provides helpful historical background of the origins and evolution of the Black Muslim worldview, linking useful insights on the emergence of religion in general to Islamic and Christian belief in Africa and in the United States. But his study does not engage the contradictions of belief and ambiguities of emotion that characterized Malcolm's moral life. In fairness to Lomax, this study was not his final word on Malcolm. But his later comparative biography of Malcolm and King is more striking for its compelling personal insight into two tragic, heroic men than for its comprehension of the constellation of cultural factors that shaped their lives.

Cone's *Martin and Malcolm and America,* on the contrary, is useful precisely because it explores the cultural, racial, and religious roots of Malcolm's public moral thought.[27] Cone, the widely acknowledged founder of black theology, has been significantly influenced by both King and Malcolm, and his book is a public acknowledgment of intellectual debt and personal inspiration. In chapters devoted to the impact of Malcolm's northern ghetto origins on his later thought, the content of his social vision, and the nature of his mature reflections on American society and black political activity, Cone discusses Malcolm's understanding of racial oppression, social justice, black unity, self-love, separatism, and self-defense that in the main constituted his vision of black nationalism.[28]

Cone performs a valuable service by shedding light on Malcolm's religious faith and then linking that faith to his social ideals and public moral vision, recognizing that his faith "was marginal not only in America as a whole but in the African-American community itself."[29] Cone covers familiar ground in his exposition of Malcolm's views on white Americans, black Christianity, and the religious and moral virtues of Elijah Muhammad's Black Muslim faith. But he also manages to show how Malcolm's withering criticisms of race anticipated "the rise of black liberation theology in the United States and South Africa and other expressions of liberation theology in the Third World."[30]

The most prominent feature of Cone's book is its comparative framework, paralleling and opposing two seminal influences on late-twentieth-century American culture. It is just this presumption—that Malcolm and Martin represented two contradictory, if not mutually exclusive, ideological options available to blacks in combating the absurdity of white racism—that generates interest in Cone's book, and in Lomax's *To Kill a Black Man*.[31] But is this presumption accurate?

As with all strictly imagined oppositions, an either–or division does not capture what Ralph Ellison termed the "beautiful and confounding complexities of Afro-American culture."[32] Nor does a rigid dualism account for the fashion in which even sharp ideological differences depend on some common intellectual ground to make disagreement plausible. For instance, the acrimonious ideological schism between Booker T. Washington and W. E. B. Du Bois drew energy from a common agreement that something must be done about the black cultural condition, that intellectual investigation must be wed to cultural and political activity in addressing the various problems of black culture, and that varying degrees of white support were crucial to the attainment of concrete freedom for black Americans.[33] Although Washington is characterized as an "accommodationist" and Du Bois as a "Pan-African nationalist," they were complex human beings whose political activity and social thought were more than the sum of their parts.

The comparative analysis of King and Malcolm sheds light on the strengths and weaknesses of the public-moralist approach to Malcolm's life and career. By comparing the two defining figures of twentieth-century black public morality, we are allowed to grasp the experienced, lived-out distinctions between King's and Malcolm's approaches to racial reform and revolution. Because King and Malcolm represent as well major tendencies in historic black ideological warfare against white racism, their lives and thought are useful examples of the social strategies, civil rebellion, religious resources, and psychic maneuvers adopted by diverse black movements for liberation within American society.

The challenge to the public-moralist approach is to probe the sort of tensions between King and Malcolm that remain largely unexplored by other views of either figure. For instance, it is the presence of class differences within black life that bestowed particular meanings on King's and Malcolm's leadership. Such differences shaped the styles each leader adapted in voicing the grievances of his constituency—for King, a guilt-laden, upwardly mobile, and ever-expanding black middle class; for Malcolm, an ever-widening, trouble-prone, and rigidly oppressed black ghetto poor. These differences reflect deep and abiding schisms within African-American life that challenge facile or pedestrian interpretations of black leaders, inviting instead complex theoretical analyses of their public moral language and behavior.

The comparison of King and Malcolm may also, ironically, void the self-critical dimensions of the public-moralist perspective, causing its proponents to leave unaddressed, for instance, the shortcomings of a sexual hierarchy of social criticism in black life. Although Cone is critical of Malcolm's and Martin's failures of sight and sense on gender issues, more is demanded. What we need is an explanation of how intellectuals and leaders within vibrant traditions of black social criticism seem, with notable exception, unwilling or unable to include gender difference as a keyword in their public-moralist vocabulary. A comparative analysis of King and Malcolm may point out how *they* did not take gender difference seriously, but it does not explain how the public-moralist traditions in which they participated either enabled or prevented them from doing so.

By gaining such knowledge, we could determine if their beliefs were representative of their traditions, or if other participants (for example, Douglass and Du Bois, who held more enlightened views on gender) provide alternative perspectives from which to criticize Malcolm and Martin without resorting to the fingerpointing that derives from the clear advantage of historical hindsight.

As Cone makes clear, Malcolm and Martin were complex political actors whose thought derived from venerable traditions of response to American racism, usefully characterized as nationalism and integrationism. But as Cone also points out, the rhetoric of these two traditions has been employed to express complex beliefs, and black leaders and intellectuals have often combined them in their struggles against slavery and other forms of racial oppression.

Lomax, by comparison, more rigidly employs these figures to "examine the issues of 'integration versus separation,' 'violence versus nonviolence,' 'the relevance of the Christian ethic to modern life,' and the question 'can American institutions as now constructed activate the self-corrective power that is the basic prerequisite for racial harmony?' ''[34] Lomax is most critical of Malcolm, leading one commentator to suggest that Lomax's assessment of Malcolm betrayed their friendship.[35] Lomax points out the wrongheadedness of Malcolm's advocacy of violence, the contradictions of his ideological absolutism, and the limitations of his imprecisely formulated organizational plans in his last year. His criticisms of King, however, are mostly framed as the miscalculations of strategy and the failure of white people to justify King's belief in them. Lomax's vision of Malcolm loses sight of the formidable forces that were arrayed against him, and the common moral worldviews occupied by King and his white oppressors, which made King's philosophical inclinations seem natural and legitimate, and Malcolm's, by that measure, foreign and unacceptable. One result of Lomax's lack of appreciation for this difference is his failure to explore King's challenge to capitalism, a challenge that distinguished King from Malcolm for most of Malcolm's career.

Another problem is that we fail to gain a more profitable view of Malcolm's real achievements, overlooking the strengths and weaknesses of the moral tradition in which he notably participated. Malcolm was, perhaps, the living indictment of a white American moral worldview. But his career was the first fruit as well of something more radical: an alternative racial cosmos where existing moral principles are viewed as the naked justification of power and thought to be useless in illumining or judging the propositions of an authentically black ethical worldview. Not only did Malcolm call for the rejection of particular incarnations of moral viewpoints that have failed to live up to their own best potential meanings (a strategy King employed to brilliant effect), but, given how American morality is indivisible from the network of intellectual arguments that support and justify it, he argued for the rejection of American public morality itself. Malcolm lived against the fundamental premises of American public moral judgment: that innocence and corruption are on a continuum, that justice and injustice are on a scale, and that proper moral choices reflect right decisions made between good and evil within the given moral outlook.

Malcolm's black Islamic moral criticism posed a significant challenge to its black Christian counterpart, which has enjoyed a central place in African-American culture. Malcolm challenged an assumption held by the most prominent black Christian public moralists: that the social structure of American society should be rearranged, but not reconstructed. Consequently, Malcolm focused a harshly critical light on the very possibility of interracial cooperation, common moral vision, and social coexistence.

A powerful vision of Malcolm as a public moralist can be seen in Goldman's *The Death and Life of Malcolm X*. Goldman captures with eloquence and imagination the Brobdingnagian forces of white racial oppression that made life hell for northern poor blacks, and the Lilliputian psychic resources apparently at their disposal before Malcolm's oversized and defiant rhetoric rallied black rage and anger to their defense. Goldman's Malcolm is one whose "life was itself an accusation—a passage to the ninth circle of that black man's hell and back—and the real meaning of his ministry, in and out of the Nation of Islam, was to deliver that accusation to us." Malcolm was a "witness for the prosecution" of white injustice, a "public moralist." With each aspect of Malcolm's life that he treats—whether his anticipation of Black Power or his capitulation to standards of moral evaluation rooted in the white society he so vigorously despised—Goldman's narrative skillfully defends the central proposition of Malcolm's prophetic public moral vocation.[36]

Goldman's book is focused on Malcolm's last years before his break with Muhammad, and tracks Malcolm's transformation after Mecca. Goldman contends that this transformation occurred as process, not revelation, and that it ran over weeks and months of trial and error, discovery, disappointment. Additionally, Goldman sifts through the conflicting evidence of Malcolm's assassination.[37] Goldman maintains that only one of Malcolm's three convicted and imprisoned assassins is justly jailed, and that two other murderers remain free.[38] Goldman says about Malcolm's Organization of Afro-American Unity (OAAU), which he founded in his last year, that its "greatest single asset was its star: its fatal flaw was that it was constructed specifically as a star vehicle for a man who didn't have the time to invest in making it go."[39]

When it was written in 1973, and revised in 1979, Goldman's was the only full-length biography of Malcolm besides Lomax's *To Kill a Black Man*. The virtue of Goldman's book is that it taps into the sense of immediacy that drives Lomax's book, while also featuring independent investigation of Malcolm's life through more than a hundred interviews with Malcolm himself. Goldman's treatment of Malcolm also raises a question that I will more completely address later: Can a white intellectual understand and explain black experience? Goldman's book helps expose the cultural roots and religious expression of Elijah Muhammad's social theodicy, an argument Malcolm took up and defended with exemplary passion and fidelity. He describes Malcolm's public moral mission to proclaim judgment on white America with the same kind of insight and clarity that characterized many of Malcolm's public declarations.

Explaining Malcolm as a public moralist moves admirably beyond heroic reconstruction to critical appreciation. The significance of such an approach is its insistence on viewing Malcolm as a critical figure in the development of black nationalist repudiations of white cultural traditions, economic practices, and religious institutions. And yet, unlike hero worshipers who present treatments of Malcolm's meaning, the authors who examine the moral dimensions of Malcolm's public ministry are unafraid to be critical of his ideological blindnesses, his strategic weakness, his organizational limitations, and his sometimes bristling moral contradictions.

But if they display an avidity, and aptitude, for portraying Malcolm's moral dimensions and the forces that made his vision necessary, Malcolm's public-moralist interpreters have not as convincingly depicted the forces that make public morality possible. The public-moralist approach is almost by definition limited to explaining Malcolm in terms of the broad shifts and realignment of contours created within the logic of American morality itself, rarely asking whether public moral proclamation and action are the best means of effecting social revolution. This approach largely ignores the hints of rebellion against capitalist domination contained in Malcolm's latest speeches, blurring as well a focus on King's mature beliefs that American society was "sick" and in need of a "reconstruction of the entire society, a revolution of values."[40]

This approach also fails to place Malcolm in the intricate nexus of social and political forces that shaped his career as a religious militant and a revolutionary black nationalist. It does not adequately convey the mammoth scope of economic and cultural forces that converged during the 1940s, 1950s, and 1960s, not only shaping the expression of racial domination, but influencing as well patterns of class antagonism and gender oppression. As Clayborne Carson argues in his splendid introduction to the FBI files on Malcolm X, most writings have failed to "study him within the context of American racial politics during the 1950s and 1960s."[41] According to Carson, the files track Malcolm's growth from the "narrowly religious perspective of the Nation of Islam toward a broader Pan-Africanist worldview," shed light on his religious and political views and the degree to which they "threatened the American state," and "clarif[y] his role in modern African-American politics."[42]

Moreover, the story of Malcolm X and the black revolution he sought to effect is also the story of how such social aspirations were shaped by the advent of nuclear holocaust in the mid-1940s (altering American ideals of social stability and communal life expectation), the repression of dissident speech in the 1950s under the banner of McCarthyism, and the economic boom of the mid-1960s that contrasted starkly to shrinking resources for the black poor. A refined social history not only accents such features, but provides as well a complex portrait of Malcolm's philosophical and political goals, and the myriad factors that drove or denied their achievement.

Malcolm's most radical and original contribution rested in reconceiving the possibility of being a worthful black human being in what he deemed a wicked white world. He saw black racial debasement as the core of an alternative moral sphere that was justified for no other reason than its abuse and attack by white Americans. To understand and explain Malcolm, however, we must wedge beneath the influences that determined his career in learning how his public moral vocation was both necessary and possible.

Psychobiography and the Forces of History

If the task of biography is to help readers understand human action, the purpose of psychobiography is to probe the relationship between psychic motivation, personal behavior, and social activity in explaining human achievement and failure. The project to connect psychology and biography grows out of a well-established quest to merge various schools of psychological theory with other intellectual disciplines, resulting in ethnopsychiatry, psychohistory, social psychology, and psychoanalytic approaches to philosophy.[43]

Behind the turn toward psychology and social theory by biographers is a desire to take advantage of the insight yielded from attempts to correlate or synthesize the largely incompatible worlds of psychoanalysis and Marxism carved out by Freud and Marx and their unwieldy legion of advocates and interpreters. If one argues, however, as Richard Lichtman does, that "the structure of the two theories makes them ultimate rivals," then, as he concedes, "priorities must be established."[44] In his analysis of the integration of psychoanalysis into Marxist theory, Lichtman argues that "working through the limitations of Freud's view makes its very significant insights available for incorporation into an expanded Marxist theory."[45]

Psychobiographers have acknowledged the intellectual difficulties to which Lichtman points while using Marxist or Freudian theory (and sometimes both) to locate and illumine gnarled areas of human experience. For instance, Erik Erikson's *Gandhi's Truth: On the Origins of Militant Nonviolence*, one of psychobiography's foundational works, weds critical analysis of its subject's cultural and intellectual roots to imaginative reflections on the sources of Gandhi's motivation, sacrifice, and spiritual achievement.[46]

As they bring together social and psychological theory in their research, psychobiographers often rupture the rules that separate academic disciplines. Then again, if the psychobiographer is ruled by rigid presuppositions and is insensitive to the subject of study, nothing can prevent the results from being fatally flat. Two recent psychobiographies of Malcolm X reveal that genre's virtues and vices.

Eugene Victor Wolfenstein's *The Victims of Democracy: Malcolm X and the Black Revolution* is a work of considerable intellectual imagination and rigorous theoretical insight.[47] It takes measure of the energies that created Malcolm and the demons that drove him. Wolfenstein assesses Malcolm's accomplishments through a theoretical lens as noteworthy for its startling clarity about Malcolm the individual as for its wide-angled view of the field of forces with which Malcolm contended during his child-hood and mature career.

Wolfenstein uses an elaborate conceptual machinery to examine how racism falsifies "the consciousness of the racially oppressed," and how racially oppressed individuals struggle to "free themselves from both the falsification of their consciousness and the racist domination of their practical activity."[48] For Wolfenstein's purpose, neither a psychoanalytic nor a Marxist theory alone could yield adequate insight because Freudianism "provides no foundation for the analysis of interests, be they individual or collective," and Marxism "provides no foundation for the analysis of desires." Therefore, a "unifying concept of human nature was required."[49]

Wolfenstein's psychobiography is especially helpful because it combines several compelling features: a historical analysis of the black (nationalist) revolutionary struggle, an insightful biographical analysis of Malcolm X's life, and an imaginative social theory that explains how a figure like Malcolm X could emerge from the womb of black struggle against American apartheid. Wolfenstein accounts for how Malcolm's childhood was affected by violent, conflicting domestic forces and describes how black culture's quest for identity at the margins of American society—especially when viewed from the even more marginal perspective of the black poor—shaped Malcolm's adolescence and young adulthood.

Wolfenstein also explores Malcolm's career as a zealous young prophet and public mouthpiece for Elijah Muhummad, revealing the psychic and social needs that Malcolm's commitments served. Wolfenstein's imaginative remapping of Malcolm's intellectual and emotional landscape marks a significant contribution as well to the history of African-American ideas, offering new ways of understanding one of the most complex figures in our nation's history.

Undoubtedly, Wolfenstein's book would have benefited from a discussion of how black religious groups provided social and moral cohesion in northern urban black communities, and from a description of their impact on Earl Little's ministry. Although Wolfenstein perceptively probes the appeal of Marcus Garvey's Universal Negro Improvement Association to blacks—and the social, psychological, and economic ground it partly shared with the Ku Klux Klan and white proletarian workers—his psychoanalytic Marxist interpretation of Earl Little and Malcolm would have been substantially enhanced by an engagement with black Protestant beliefs about the relationship between work, morality, and self-regard.[50]

Wolfenstein is often keenly insightful about black liberation movements and the forces that precipitated their eruption, but his dependence on biological definitions of race weakens his arguments.[51] The value of more complex readings of race is that they not only show how the varied meanings of racism are created in society; but prove as well that the idea of race has a cultural history.[52] More complex theories of race would permit Wolfenstein to illumine the changing intellectual and social terrain of struggle by groups that oppose the vicious meanings attributed to African-American identity by cultural racists.

In the end, Wolfenstein is too dependent on the revelations and reconstructions of self-identity that Malcolm (with Haley's assistance) achieved in his autobiography. In answering his own rhetorical questions about whether Malcolm and Haley represented Malcolm accurately, Wolfenstein says that from a "purely empirical standpoint, I believe the answer to both questions is generally affirmative."[53] The problem, of course, is that Malcolm's recollections are not without distortions. These distortions, when taken together with the book's interpretive framework, not only reveal his attempts to record his life history, but reflect as well his need to control how his life was viewed during the ideological frenzy that marked his last year. By itself, self-description is an unreliable basis for reconstructing the meaning of Malcolm's life and career. Still, Wolfenstein's work is the most sophisticated treatment to date of Malcolm's intellectual and psychological roots.

But Bruce Perry's uneven psychobiographical study, *Malcolm: The Life of a Man Who Changed Black America,* which reaches exhaustively beyond Malcolm's self-representation in his autobiography, possesses little of the psychoanalytic rigor and insight of Wolfenstein's work.[54] Although Perry unearths new information about Malcolm, he does not skillfully clarify the impact that such information should have on our understanding of Malcolm. The volume renders Malcolm smaller than life.

In Perry's estimation, Malcolm's childhood holds the interpretive key to understanding his mature career as a black leader: Malcolm's "war against the white power structure evolved from the same inner needs that had spawned earlier rebellions against his teachers, the law, established religion, and other symbols of authority."[55] Perry's picture of Malcolm's family is one of unremitting violence, criminality, and pathology. The mature Malcolm is equally tragic: a man of looming greatness whose self-destruction "contributed to his premature death."[56] It is precisely here that Perry's psychobiography folds in on itself, its rough edges puncturing the center of its explanatory purpose. It is not that psychobiography cannot remark on the unraveling of domestic relations that weave together important threads of personal identity, threads that are also woven into adolescent and adult behavior. But Perry has a penchant for explaining complex psychic forces—and the social conditions that influence their makeup—in simplistic terms and tabloid-like arguments.

Still, Perry's new information about Malcolm is occasionally revealing, though some of the claims he extracts from this information are more dubious than others. When, for instance, Perry addresses areas of Malcolm's life that can be factually verified, he is on solid ground. By simply checking Malcolm's school records Perry proves that, contrary to his autobiography, Malcolm was not expelled from West Junior High School but actually completed the seventh grade in 1939. And by interviewing several family members, Perry establishes that neither Malcolm's half-sister Ella nor his father, Earl, were, as Malcolm contended, "jet black," a claim Perry views as Malcolm's way of equating "blackness and the strength his light-skinned mother had lacked."[57] Despite Malcolm X's assertion of close friendships with Lionel Hampton, Sonny Greer, and Cootie Williams during his hustling days, Perry's interviews show that the "closeness Malcolm described was as fictitious as the closeness he said he had shared with the members of his own family."[58]

But when Perry addresses aspects of Malcolm's experience that invite close argument and analytical interpretation, he is on shakier ground. At this juncture, Perry displays an insensitivity to African-American life and an ignorance about black intellectual traditions that weaken his book. For instance, Perry depicts Malcolm's travels to Africa—partially in an attempt to expand his organization's political and financial base, but also to express his increasingly international social vision—as intended solely to fund his fledgling organization. Perry also draws questionable parallels between the cloudy events surrounding a fire at Malcolm's family farm during his early childhood in 1929 (which Perry concludes points to arson by Earl Little) and the fire at Malcolm's New York house after his dispute with Nation of Islam officials over ownership rights.

A major example of the limitation of Perry's psycho-biographical approach is his treatment of Malcolm's alleged homosexual activity, both as an experimenting adolescent and as a hustling, income-seeking young adult. Perry's remarks are more striking for the narrow assumptions that underlie his interpretations than for their potential to dismantle the quintessential symbol of African-American manhood. If Malcolm did have homosexual relations, they might serve Perry as a powerful tool of interpretation to expose the tangled cultural roots of black machismo, and to help him explain the cruel varieties of homophobia that afflict black communities. A complex understanding of black sexual politics challenges a psychology of masculinity that views "male" as a homogeneous, natural, and universally understood identity. A complex understanding of masculinity maintains that male identity is also significantly affected by ethnic, racial, economic, and sexual differences.

But Perry's framework of interpretation cannot assimilate the information his research has unearthed. Although the masculinist psychology that chokes much of black leadership culture needs to be forcefully criticized, Perry's observations do not suffice. Because he displays neither sensitivity to nor knowledge about complex black cultural beliefs regarding gender and sexual difference, Perry's portrait of Malcolm's sex life forms a rhetorical low blow, simply reinforcing a line of attack against an already sexually demonized black leadership culture.

The power of psychobiography in discussing black leaders is its potential to shed light on its subjects in a manner that traditional biography fails to achieve. African-American cultural studies, which has traditionally made little use of psychoanalytic theory, has sacrificed the insights such an undertaking might offer while avoiding the pitfalls of psychological explanations of human motivation. After all, psychobiography is also prone to overreach its capacity to explain.

In some ways, the psychobiographer's quest for (in this case) the "real Malcolm" presumes that human experience is objective and that truth is produced by explaining the relation between human action and psychic motivation. Such an approach may seduce psychobiographers into believing that they are gaining access to the static, internal psychic reality of a historical figure. Often such access is wrongly believed to be separate from the methods of investigation psychobiographers employ, and from the aims and presumptions, as well as the biases and intellectual limitations, that influence their work.

Because both Wolfenstein and Perry (like Goldman) are white, their psychobiographies in particular raise suspicion about the ability of white intellectuals to interpret black experience. Although such speculation is rarely systematically examined, it surfaces as both healthy skepticism and debilitating paranoia in the informal debates that abound in a variety of black intellectual circles. Such debates reflect two crucial tensions generated by psychobiographical explanations of black leaders by white authors: that such explanations reflect insensitivity to black culture, and that white proponents of psychobiographical analysis are incompetent to assess black life adequately. Several factors are at the base of such conclusions.

First is the racist history that has affected every tradition of American scholarship and that has obscured, erased, or distorted accounts of the culture and history of African-Americans.[59] Given this history (and the strong currents of anti-intellectualism that flood most segments of American culture), suspicion of certain forms of critical intellectual activity survive in many segments of black culture. Also, black intellectuals have experienced enormous difficulty in securing adequate cultural and financial support to develop self-sustaining traditions of scholarly investigation and communities of intellectual inquiry.[60]

For example, from its birth in the womb of political protest during the late 1960s and early 1970s, black studies has been largely stigmatized and usually underfunded. Perhaps the principal reasons for this are the beliefs held by many whites (and some blacks) that, first, black scholars should master nonblack subjects, and second, that black studies is intellectually worthless. Ironically, once the more than 200 black studies programs in American colleges and universities became established, many white academics became convinced that blacks are capable of studying only "black" subjects.

At the same time, black studies experienced a new "invasion" by white intellectuals. This new invasion—mimicking earlier patterns of white scholarship on black life even as most black scholars were prevented from being published—provoked resentment from black scholars.[61] The resentment hinged on the difficulty black scholars experienced in securing appointments in most academic fields beyond black studies. Black scholars were also skeptical of the intellectual assumptions and political agendas of white scholars, especially because there was strong precedence for many white scholars to distort black culture in their work by either exoticizing or demonizing its expression. Black intellectual skeptics opposed to white interpretations of black culture and figures employ a variety of arguments in their defense.

Many black intellectuals contend that black experience is unique and can be understood, described, and explained only by blacks. Unquestionably, African-American history produces cultural and personal experiences that are distinct, even singular. But the *historical* character of such experiences makes them theoretically accessible to any interpreter who has a broad knowledge of African-American intellectual traditions, a balanced and sensible approach to black culture, and the same skills of rational argumentation and scholarly inquiry required in other fields of study.

There is no special status of being that derives from black cultural or historical experience that grants black interpreters an automatically superior understanding of black cultural meanings. This same principle allows black scholars to interpret Shakespeare, study Heisenberg's uncertainty principle, and master Marxist social theory. In sum, black cultural and historical experiences do not produce ideas and practices that are incapable of interpretation when the most critically judicious and culturally sensitive methods of intellectual inquiry are applied.

Many intellectuals also believe that black culture is unified and relatively homogeneous. But this contention is as misleading as the first, especially in light of black culture's wonderful complexity and radical diversity. The complexity and diversity of black culture means that a bewildering variety of opinions, beliefs, ideologies, traditions, and practices coexist, even if in a provisional sort of way. Black conservatives, scuba divers, socialists, and rock musicians come easily to mind. All these tendencies and traditions constitute and help define black culture. Given these realities, it is pointless to dismiss studies of black cultural figures *simply* because their authors are white. One must judge any work on African-American culture by standards of rigorous critical investigation while attending to both the presuppositions that ground scholarly perspectives and the biases that influence intellectual arguments.

Psychobiographies of Malcolm X's life and career represent an important advance in Malcolm studies. The crucial issue is not color, but consciousness about African-American culture, sensitivity to trends and developments in black society, knowledge of the growing literature about various dimensions of black American life, and a theoretical sophistication that artfully blends a variety of disciplinary approaches in yielding insight about a complex historic figure like Malcolm X. When psychobiography is employed in this manner, it can go a long way toward breaking new ground in understanding and explaining the life of important black figures. When it is incompetently wielded, psychobiographical analysis ends up simply projecting the psychobiographer's intellectual biases and limitations of perspective onto the historical screen of a black figure's career.

Voices in the Wilderness: Revolutionary Sparks and Malcolm's Last Year

To comprehend the full sweep of a figure's life and thought, it is necessary to place that figure's career in its cultural and historical context and view the trends and twists of thought that mark significant periods of change and development.

Such an approach may be termed a trajectory analysis because it attempts to outline the evolution of belief and thought of historic figures by matching previously held ideas to newer ones, seeking to grasp whatever continuities and departures can be discerned from such an enterprise. Trajectory analysis, then, may be a helpful way of viewing a figure such as Martin Luther King, Jr., whose career may be divided into the early optimism of civil rights ideology to the latter-day aggressive nonviolence he advocated on the eve of his assassination. It may also be enlightening when grappling with the serpentine mysteries of Malcolm's final days.

Malcolm's turbulent severance from Elijah Muhammad's psychic and world-making womb initiated yet another stage of his personal and political evolution, marking a conversion experience. On one level, Malcolm freed himself from Elijah's destructive ideological grip, shattering molds of belief and practice that were no longer useful or enabling. On another level, Malcolm's maturation and conversion were the result of his internal ideals of moral expectation, social behavior, and authentic religious belief. His conversion, though suddenly manifest, was most likely a gradual process involving both conscious acts of dissociation from the Nation of Islam and the "subconscious incubation and maturing of motives deposited by the experiences of life."[62]

Many commentators have heavily debated the precise

nature of Malcolm's transformation. Indeed, his last fifty weeks on earth form a fertile intellectual field where the seeds of speculation readily blossom into conflicting interpretations of Malcolm's meaning at the end of his life. Lomax says that Malcolm became a "lukewarm integrationist."[63] Goldman suggests that Malcolm was "improvising," that he embraced and discarded ideological options as he went along.[64] Cleage and T'Shaka hold that he remained a revolutionary black nationalist. And Cone asserts that Malcolm became an internationalist with a humanist bent.

But the most prominent and vigorous interpreters of the meaning of Malcolm's last year have been a group of intellectuals associated with the Socialist Workers Party, a Trotskyist Marxist group that took keen interest in Malcolm's post-Mecca social criticism and sponsored some of his last speeches. For the most part, their views have been articulately promoted by George Breitman, author of *The Last Year of Malcolm X: The Evolution of a Revolutionary* and editor of two volumes of Malcolm's speeches, organizational statements, and interviews during his last years: *Malcolm X Speaks: Selected Speeches and Statements* and *By Any Means Necessary: Speeches, Interviews, and a Letter, by Malcolm X*. A third volume of Malcolm's speeches, *Malcolm X: The Last Speeches*, was edited by Bruce Perry, who claimed ideological difference with the publisher.[65]

Breitman's *The Last Year of Malcolm X* is a passionately argued book that maintains Malcolm's split with Elijah took

Malcolm by surprise, making it necessary for him to gain time and experience to reconstruct his ideological beliefs and redefine his organizational orientation. Breitman divides Malcolm's independent phase into two parts: the transition period, lasting the few months between his split in March 1964 and his return from Africa at the end of May 1964; and the final period, lasting from June 1964 until his death in February 1965. Breitman maintains that in the final period, Malcolm "was on the way to a synthesis of black nationalism and socialism that would be fitting for the American scene and acceptable to the masses in the black ghetto."[66]

For Breitman's argument to be persuasive, it had to address Malcolm's continuing association with a black nationalism that effectively excluded white participation, or else show that he had developed a different understanding of black nationalism. Also, he had to prove that Malcolm's anticapitalist statements and remarks about socialism represented a coherent and systematic exposition of his beliefs as a political strategist and social critic. Breitman contends that in the final period, Malcolm made distinctions between separatism (the belief that blacks should be socially, culturally, politically, and economically separate from white society) and nationalism (the belief that blacks should control their own culture).

Malcolm's views of nationalism changed after his encounters with revolutionaries in Africa who were "white," however, and in his "Young Socialist" interview in *By Any Means Necessary,* Malcolm confessed that he had "had to do a lot of thinking and reappraising" of his definition of black nationalism.[67] Breitman argues that though he "had virtually stopped calling himself and the OAAU black nationalist," because others persisted in the practice, he accepted "its continued use in discussion and debate."[68] Malcolm said in the same interview, "I haven't been using the expression for several months."[69]

But how can Breitman then argue that Malcolm was attempting a synthesis of black nationalism and socialism if the basis for Malcolm's continued use of the phrase "black nationalism" was apparently more convenience and habit than ideological conviction? What is apparent from my reading of Malcolm's speeches is that his reconsideration of black nationalism occurred amid a radically shifting worldview that was being shaped by events unfolding on the international scene and by his broadened horizon of experience. His social and intellectual contact with activists and intellectuals from several African nations forced him to relinquish the narrow focus of his black nationalist practice and challenged him to consider restructuring his organizational base to reflect his broadened interests.

If, therefore, even Malcolm's conceptions of black nationalist strategy were undergoing profound restructuring, it is possible to say only that his revised black nationalist ideology *might* have accommodated socialist strategy. It is equally plausible to suggest that his nationalist beliefs might have collapsed altogether under the weight of apparent ideological contradictions introduced by his growing appreciation of class and economic factors in forming the lives of the black masses.[70] For the synthesis of black nationalism and socialism that Breitman asserts Malcolm was forging to have been plausible, several interrelated processes needed to be set in motion.

First, for such a synthesis to have occurred, a clear definition of the potential connection of black nationalism and socialism was needed. The second need was for a discussion of the ideological similarities and differences between the varieties of black nationalism and socialism to be joined. And the third need was for an explicit expression of the political, economic, and social interests that an allied black nationalism and socialism would mutually emphasize and embrace; the exploration of intellectual and political problems both would address; and an identification of the common enemies both would oppose. But given the existential and material matters that claimed his rapidly evaporating energy near the end of his life, Malcolm hardly had the wherewithal to perform such tasks.

Breitman also maintains that Malcolm's final period marked his maturation as "a revolutionary—increasingly anti-capitalist and pro-socialist as well as anti-imperialist," labels that Breitman acknowledges Malcolm himself never adopted.[71] Breitman reads Malcolm's two trips to Africa as a time of expansive political reeducation, when Malcolm gained insight into the progressive possibilities of socialist revolutionary practice. After his return to the United States from his second trip, Malcolm felt, Breitman says, the need to express publicly his "own anti-capitalist and pro-socialist convictions," which had "become quite strong by this time."[72] He cites interviews and speeches Malcolm made during this period to substantiate his claim, including Malcolm's speaking at the Audubon Ballroom on December 20, 1964, of how almost "every one of the countries that has gotten independence has devised some kind of socialist system, and this is no accident."[73]

Such a strategy, one that seeks to predict probable ideological and intellectual outcomes, may shed less light on Malcolm than is initially apparent. Breitman's contention that Malcolm was becoming a socialist; Cleage's that he was confused; T'Shaka's that he maintained a vigorous revolutionary black nationalist stance; and Goldman's that he was improvising can all be proclaimed and documented with varying degrees of evidence and credibility.

This is not to suggest that one view is as good as the next or that they are somehow interchangeable, because we are uncertain about Malcolm's final direction. It simply suggests that the nature of Malcolm's thought during his last year was ambiguous and that making definite judgments about his direction is impossible. In this light, trajectories say more about the ideological commitments and intellectual viewpoints of interpreters than the objective evidence evoked to substantiate claims about Malcolm's final views.

The truth is that we have only a bare-bones outline of Malcolm's emerging worldview. In "The Harlem 'Hate-Gang' Scare," contained in *Malcolm X Speaks* (and delivered during what Breitman says was Malcolm's final period), Malcolm says that during his travels he

> noticed that most of the countries that had recently emerged into independence have turned away from the so-called capitalistic system in the direction of socialism. So out of curiosity, I can't resist the temptation to do a little investigating wherever that particular philosophy happens to be in existence or an attempt is being made to bring it into existence.[74]

But at the end of his speech, in reply to a question about the kind of political and economic system that Malcolm wanted, he said, "I don't know. But I'm flexible. . . . As was stated earlier, all of the countries that are emerging today from under the shackles of colonialism are turning toward socialism."[75]

This tentativeness is characteristic of Malcolm's

speeches throughout the three collections that contain fragments of his evolving worldview, especially *Malcolm X Speaks* and *By Any Means Necessary*. Even the speeches delivered during his final period showcase a common feature: Malcolm displays sympathy for and interest in socialist philosophy without committing himself to its practice as a means of achieving liberation for African-Americans.

Malcolm confessed in the "Young Socialist" interview, "I still would be hard pressed to give a specific definition of the overall philosophy which I think is necessary for the liberation of the black people in this country."[76] Of course, as Breitman implies, Malcolm's self-description is not the only basis for drawing conclusions about his philosophy. But even empirical investigation fails to yield conclusive evidence of his social philosophy because it was in such radical transformation and flux.

Malcolm was indeed improvising from the chords of an expanded black nationalist rhetoric and an embryonic socialist criticism of capitalist civilization. Although Breitman has been maligned as a latecomer seeking to foist his ideological beliefs onto Malcolm's last days, there is precedence for Trotskyist attempts to address the problem of racism and black nationalism in the United States.[77] And the venerable black historian C. L. R. James became a Marxist, in part, by reading Trotsky's *History of the Russian Revolution*.[78] Although Malcolm consistently denounced capitalism, he did not live long enough to embrace social-ism.

The weakness of such an interpretive trajectory, then, is that it tends to demand a certainty about Malcolm that is clearly unachievable. An ideological trajectory of Malcolm's later moments is forced to bring coherence to fragments of political speech more than systematic social thought, to ex-aggerate moments of highly suggestive ideological gestures rather than substantive political activity, and to focus on slices of organizational breakthrough instead of the complex integrative activity envisioned for the OAAU. In the end, it is apparent that Malcolm was rapidly revising his worldview as he experienced a personal, religious, and ideological con-version that was still transpiring when he met his brutal death.

But the thrust behind such speculation is often a focus on how Malcolm attempted to shape the cultural forces of his time through the agency of moral rhetoric, social criti-cism, and prophetic declaration. Just as important, but often neglected in such analyses, is an account of how Malcolm was shaped by his times, of how he was the peculiar and particular creation of black cultural forces and American social practices. Armed with such an understanding, the fo-cus on Malcolm's last year would be shifted away from sim-ply determining what he said and did to determining how we should use his example to respond to our current cul-tural and national crises.

In the Prison of Prisms:
The Future of Malcolm's Past

The literature on Malcolm X is certain to swell with the renewed cultural interest in his life. And although the particular incarnations of the approaches I have detailed may fade from intellectual view or cultural vogue, the ideological commitments, methodological procedures, historical perspectives, cultural assumptions, religious beliefs, and philosophical presuppositions they employ will most assuredly be expressed in one form or another in future treatments of his life and thought.[79]

The canonization of Malcolm will undoubtedly continue. Romantic and celebratory treatments of his social action and revolutionary rhetoric will issue forth from black intellectuals, activists, and cultural artists. This is especially true in the independent black press, where Malcolm's memory has been heroically kept alive in books, pamphlets, and magazines, even as his presence receded from wide visibility and celebration before his recent revival. The independent black press preserves and circulates cultural beliefs, intellectual arguments, and racial wisdom among black folk away from the omniscient eye and acceptance of mainstream publishing.

Shahrazad Ali's controversial book, *The Blackman's Guide to Understanding the Blackwoman,* for instance, sold hundreds of thousands of copies without receiving much attention from mainstream newspapers, magazines, or journals. The mainstream press often overlooked Malcolm's contributions, but black publications like *The Amsterdam News, The Afro-American, Bilalian News,* and *Black News* scrupulously recorded his public career. The black independent press, in alliance with various black nationalist groups throughout the country that have maintained Malcolm's heroic stature from the time of his assassination, is a crucial force in Malcolm's ongoing celebration. Such treatments of his legacy will most likely be employed by these groups to actively resist Malcolm's symbolic manipulation by what they understand to be the forces of cultural racism, state domination, commodification, and especially religious brainwashing that Malcolm detested and opposed.

The enormous influence of the culture of hip-hop on black youth, coupled with the resurgence of black cultural nationalism among powerful subcultures within African-America, suggests that Malcolm's heroic example will continue to be emulated and proclaimed. The stakes of hero worship are raised when considering the resurgent racism of American society and the increased personal and social desperation among the constituency for whom Malcolm eloquently argued, the black ghetto poor. Heightened racial antipathy in cultural institutions such as universities and businesses, and escalated attacks on black cultural figures, ideas, and movements, precipitate the celebration of figures who embody the strongest gestures of resistance to white racism.

Moreover, the destructive effects of gentrification, economic crisis, and social dislocation; the expansion of corporate privilege; and the development of underground political economies—along with the violence and criminality they breed—means that Malcolm is even more a precious symbol of the self-discipline, self-esteem, and moral leadership necessary to combat the spiritual and economic corruption of poor black communities. With their efforts to situate him among the truly great in African-American history, hero worshipers' discussion of Malcolm will be of important but limited value in critically investigating his revolutionary speech, thought, and action.

Malcolm's weaknesses and strengths must be rigorously examined if we are to have a richly hued picture of one of the most intriguing figures of twentieth-century public life in the United States. Malcolm's past is not yet settled, savaged as it has been in the embrace of unprincipled denigrators while being equally smothered in the well-meaning grip of romantic and uncritical loyalists. He deserves what every towering and seminal figure in history should receive: comprehensive and critical examination of what he said and did so that his life and thought will be useful to future generations of peoples in struggle around the globe.

MALCOLM X'S INTELLECTUAL LEGACY

As the cadre of Malcolm scholars expands, Malcolm's relation to black nationalism must be explored, especially because its themes and goals occupied so much of his life and thought. I will now turn to a discussion of how Malcolm's renewed popularity is wedded to a resurgence of black nationalist sentiment. The strengths of black nationalism, and its limitations and contradictions as well, serve to magnify Malcolm's achievements and failures alike.

PART II
MALCOLM X IN CONTEMPORARY SOCIETY

X

Everything we dont understand
      is explained
      in Art
      The Sun
      beats inside us
      The Spirit courses in and out

A circling transbluesency
      pumping Detroit Red inside, deep thru us
      like a Sea
            & who calls us bitter
            has bitten us
            & from that wound
                 pours Malcolm

                        Amiri Baraka,
                            "Little
                              by
                              Little"

3
MALCOLM X AND
THE RESURGENCE OF
BLACK NATIONALISM

> And in my opinion the young generation of whites, blacks, browns, whatever else there is, you're living at a time of extremism, a time of revolution, a time when there's got to be a change. People in power have misused it, and now there has to be a change and a better world has to be built, and the only way it's going to be built is with extreme methods. I for one will join in with anyone, I don't care what color you are, as long as you want to change this miserable condition that exists on this earth.
>
> Malcolm X, in *By Any Means Necessary: Speeches, Interviews, and a Letter*, by Malcolm X

Because Malcolm X for the duration of his life and most of his death occupied the shadowy periphery of black cultural politics—subsisting as the suppressed premise of the logic of black bourgeois resistance to racism—his reemergence as a cultural hero is something of a paradox. His newly acclaimed status is indivisible from the renaissance of black nationalism and owes as much to his overhauled heroism as to his commodification by black and white cultural entrepreneurs. I will explore Malcolm's brand of black nationalism while evaluating his use as a powerful icon in contemporary black nationalism. I will end with a brief reflection on his possible use in a program of progressive black politics.

Given this nation's racist legacy, it is no surprise that black folk have at every crucial juncture of their history in the United States expressed nationalist sentiment.[1] The peculiar social, economic, and political constraints of oppression, stretching from slavery to the present day, have always precipitated varying degrees of resistance, revolt, rebellion, or resentment from African-Americans. If nationalism is viewed as an attempt to establish and maintain a nation's identity, growing out of circumstances of social and cultural conflict, then black nationalism is a response of racial solidarity to the divisive practices of white supremacist nationalism.

Black nationalism has also been viewed as a response to the erosion of communal identity and the eradication of collective self-determination under slavery, and as a strategy to combat the destructive cultural effects resulting from the rejection of fragile black political liberties after Emancipation and Reconstruction. Black nationalism was often an expression of healthy self-regard in a legal and social climate that reinforced black Americans' inferior political status. Unlike many other expressions of nationalism, however, black nationalism was coerced from the beginning into a parasitic relationship to American culture. This confounding irony of black nationalist discourse and practice haunts it to this day.

Black nationalism is often contrasted to liberal integrationist ideology. Liberal integrationists believe that the goal of African-American struggles for liberation ought to be the inclusion of blacks in the larger compass of American social, political, and economic privilege, while maintaining a distinct appreciation for African-American culture. In its extreme expression, however, liberal integrationist ideology acquires a bland assimilationist emphasis. Racial assimilationists promote the uncritical adoption by blacks of the norms of civility, education, and culture nurtured in mainstream white American culture. Although overly sharp distinctions between forms of nationalism and integrationism are problematic (the two ideologies often coexist in a figure's thought or at different periods in an institution's or organization's life), comparing them can be helpful in capturing the two primary ideological thrusts in African-American communities.

The most prominent recent phase of black nationalist activity, prior to its contemporary resurgence, lasted from 1965 until 1973, from the emergence of Stokely Carmichael as leader of the Black Power movement, until the demise of the Black Panthers.[2] This period saw major black organizations denying whites participation in radical civil rights organizations like the Student Nonviolent Coordinating

Committee, the advocacy by black nationalist leaders of armed self-defense against racist state repression in the form of the police and the National Guard, the end of the powerful leadership of Malcolm X with his assassination in 1965, the bold articulation of black theology from James Cone in 1969, and the revolutionary insurgence planned and partially implemented by the Black Panthers.[3]

The cultural rebirth of Malcolm X, then, is the remarkable result of complex forces converging to lift him from his violent death in 1965. His heroic status hinges partially on the broad, if belated, appeal of his variety of black nationalism to Americans who, when he lived, either ignored or despised him. But Malcolm's appeal is strongest among black youth between the ages of fifteen and twenty-four, who find in him a figure of epic racial achievement.[4]

Rap culture, especially, has had a decisive influence in promoting Malcolm as a cultural hero. Because of the issues it addresses, and the often militant viewpoints it espouses, rap has often served as the popular cultural elaboration of certain features of Malcolm's legacy. The obstacles that rap has overcome in establishing itself as a mainstay of American popular culture—connected primarily to its style of expression and its themes—provide a natural link between Malcolm's radical social vocation and aspects of black youth culture. The similarity between aspects of hip-hop culture and Malcolm's public career (for example, charges of violence, the problems associated with expressing black rage and experiences in the ghetto, the celebration of black pride and historical memory) prods rappers to take the lead in asserting Malcolm's heroism for contemporary black America.

Rap music originated in the Bronx over a decade ago as urban teens experimented with various forms of cultural expression, from graffiti art to break dancing, creating art in the midst of the cultural and political invisibility to which they had been relegated. Rap was initially popular among black teens because of its staccato beats, its driving, lancing rhythms, and its hip lyrics, reflecting its origins in their world, a world that is increasingly an odyssey—through the terror of ghetto gangs, drugs, violence, and racism—in search of an authentic personal identity and legitimate social standing. The seemingly endless obstacles that frustrate this search, together with the humor, nonsense, and latent absurdity of some forms of urban life, provide the content of many rap songs.

Hip-hop began as an underground phenomenon, with artists such as Busy Bee, DJ Kool Hurk, Funky 4 Plus 1, Kurtis Blow, Kool Moe Dee, Afrika Bambaata, Cold Rush Brothers, and Grandmaster Melle Mel producing cassette tapes of their verbal play and distributing them from the trunks of cars, on street corners, and at neighborhood parties. In rap's initial phase, hip-hop artists usually applied their rhythmic skills by inserting words over music borrowed from popular 1970s R&B songs. For instance, hip-hop's breakthrough song, "Rapper's Delight," featured words added to the hit "Good Times," originally recorded by the R&B group Chic.

As rap evolved, it has largely proved to be a flexible musical form that experiments widely in order to reflect the varied visions of its creators. And like most black music before it, rap has escaped classification and ghettoization as a transient "black" fad and garnered mainstream attention. The fate of rap as a "legitimate" contender for mainstream acceptance was tied initially to the fortunes of the rap group Run-D.M.C. It produced the first rap album to be certified gold (500,000 copies sold), the first rap song to be featured on MTV, and the first rap album—"Raising Hell"—to go triple platinum (3 million copies sold). Run-D.M.C.'s crossover appeal was secured with its rap version of the white rock group Aerosmith's 1970s song "Walk This Way."

Since Run-D.M.C.'s epochal success, rap has exploded all previous predictions of its cultural and commercial appeal, nearly becoming a billion-dollar industry. And the recent rise to popularity of the controversial gangsta' rap—in which rappers employ guns, violence, and drugs as metaphors for cultural creativity, personal agency, and social criticism—has only increased the visibility and demonization of black youth culture. From socially conscious rap to hardcore hip-hop, from pop rappers to black nationalist groups, rap has easily become, with country music, the most popular form of musical expression during the 1990s.

Malcolm X and the Resurgence of Black Nationalism

One of the most obvious and starkly compelling features of rap culture is its form, drawing from an oral tradition with deep roots in African-American culture.[5] The tradition is one that Malcolm brilliantly participated in, relishing his capacity to verbally outfox his opponents with a well-placed word or a cleverly engineered rebuke. His broad familiarity with the devices of African-American oral culture—the saucy put-down, the feigned agreement turned to oppositional advantage, the hyperbolic expression generously employed to make a point, the fetish for powerful metaphor—marks his public rhetoric.

The hip-hop generation has appropriated Malcolm with unequaled passion, pushed along by the same affection for the word that drove him to read voraciously and speak with eloquence. Malcolm is the rap revolution's rhetorician of choice, his words forming the ideological framework for authentic black consciousness.[6] His verbal ferocity has been combined with the rhythms of James Brown and George Clinton, the three figures forming a trio of griots dispensing cultural wisdom harnessed to polyrhythmic beats.

Malcolm's public career, too, is the powerful if perplexing story of a series of personal and intellectual changes, a constellation of complicated and sometimes conflicting identities, that mark his evolution of thought, foreshadowing the perennial transformations of style and theme that characterize contemporary hip-hop. As rapper Michael Franti, says:

> The thing I gained from him is not his symbol as a militant, but his ongoing examination of his life and how he was able to think critically about himself and grow and change as he encountered new information. That's where I feel that we gain strength, through constantly conquering our own shortcomings, and questioning our beliefs.[7]

Furthermore, as in Malcolm's public rallies—which focused black rage on suitable targets, especially black bourgeois liberal leaders and white racists—the rap concert encourages the explicit articulation of black anger in public. And like the misconceptions that often prevailed about Malcolm's provocative statements about self-defense, perceptions about the automatic or inevitable link between rap and violence are often grounded in ignorance rather than critical investigation of hip-hop's words or deeds.

Because Malcolm, too, addressed with unexcelled clarity and moral suasion the predicament of the ghetto poor, he is a natural icon for rap culture. As rappers Ultramagnetic M.C.'s state:

> Everybody still listens to Malcolm X. When he talks you can't walk away. The thing about X is that he attracted and still attracts the people who have given up and lives [sic] recklessly—the crowd that just don't care what's

going on. Making a difference in these people's lives is truly the essence of Malcolm X.[8]

Rappers often point as well to Malcolm X's phrase "no sellout, no sellout, no sellout" as the touchstone of a black cultural consciousness intent on preserving the authenticity of black cultural expressions, and as the basis for a true black nationalism.[9] But what precisely about Malcolm's black nationalist beliefs is the basis of his revived American heroism, especially for black youth?

Malcolm's defiant expression of black rage has won him a new hearing among a generation of black youth whose embattled social status due to a brutally resurgent racism makes them sympathetic to his fiery, often angry rhetoric. Malcolm's take-no-prisoners approach to racial crisis appeals to young blacks disaffected from white society and alienated from older black generations whose contained style of revolt owes more to Martin Luther King, Jr.,'s nonviolent philosophy than to Malcolm's advocacy of self-defense.

Moreover, Malcolm's expression of black rage—which, by his own confession, tapped a vulnerability even in King—has been adopted by participants in the culture of hip-hop, who often reflect Malcolm's militant posture. These artists, as do many of their black peers, find in Malcolm's uncompromising rhetoric the confirmation of their instincts about the "permanence of American racism."[10] Also, Malcolm's ability to say out loud what many blacks could say only privately endeared him to blacks when he was alive, and explains his appeal to youth seeking an explicit articulation of anger at American racism and injustice.

Another feature of Malcolm's nationalism has cemented his heroic status among young blacks: his withering indictments of the limitations of black bourgeois liberalism, expressed most clearly in the civil rights protest against white racial dominance. Malcolm showed little tolerance for the strategies, tactics, and philosophy of nonviolence that were central to the civil rights movement led by Martin Luther King, Jr.[11] Further, King's limited successes in reaching those most severely punished by poverty only reinforced the value of Malcolm's criticism of civil rights ideology.

Malcolm's pointed denunciations of black liberal protest against white racism hinged on the belief that black people should maintain independence from the very people who had helped oppress them—white people. Black bourgeois liberal protest encouraged white cooperation in the struggle to secure the fragile gains for which civil rights groups aimed in their quest for social justice. As one rap group illumines Malcolm's appeal: "The reason why Malcolm X has an influence on today's youth is because his influence as a leader was certainly equal, if not better than Dr. Martin Luther King. Everybody still listens to Malcolm X."[12] Another rap group believes that the "legacy of Mal-

colm X is to provide a clear counterpoint to the non-violent/
passive resistance theme presented by Dr. Martin Luther
King, Jr."[13] Malcolm's heroic appeal as a critic of black bourgeois protest of white racism is summarized by C. Eric Lin-
coln, who contends that the source of Malcolm's undying
magnetism

> lies in the simple fact that we have not yet overcome.
> . . . For many of the kids in the ghetto we are right back
> where we were. The few advances that have been
> made have not reached them. So if we didn't make it
> with King, what have we to lose? We might as well
> make it with Malcolm.[14]

Malcolm's black nationalist ideology expressed an alternative black spirituality and religious worldview that
provided bold relief to the ethic of love advocated in black
Christian conceptions of social protest. Although this com-
ponent of his thinking is linked to Malcolm's denunciation
of black liberal protest philosophies and strategies, his alternative black spirituality was rooted in the religious
worldview of the Nation of Islam and promoted a black public theodicy that demonized whites as unquestionably evil.[15]
And although Malcolm's understanding of white racism
was rooted in a theological vision that lent religious significance to the unequal relationship between whites and
blacks, his colorful articulation of his beliefs in his public
addresses forged the expression of a black public theodicy

with which even secular or non-Muslim blacks could identify.

Central to Malcolm's alternative black spirituality was his rejection of the belief that black people should redeem white people through black bloodshed, sacrifice, and suffering. "We don't believe that Afro-Americans should be victims any longer," he said. "We believe that bloodshed is a two-way street."[16] He also contended that not "a single white person in America would sit idly by and let someone do to him what we Black men have been letting others do to us."[17]

Malcolm's theological premises—the underpinning of his black public theodicy—forced him to the conclusion that white violence must be met with intelligent opposition and committed resistance, even if potentially violent means must be adopted in self-defense against white racism. Although Malcolm would near the end of his life alter his views and concede the humanity of whites and their potential for assistance, he maintained a strong philosophical commitment to proclaiming the evil of white racism and to detailing its lethal consequences in poor black communities.

A fundamental appeal of Malcolm's black nationalism, and indeed a large part of the cultural crisis that has precipitated Malcolm's mythic return, is rooted in a characteristic quest in black America: the search for a secure and empowering racial identity. That quest is perennially frustrated by the demands of American culture to cleanse ethnic and